Waste & Biomass Valorization

Topical Conference at the 2011 AIChE Spring Meeting and 7th Global Congress on Process Safety

Chicago, Illinois, USA
13-17 March 2011

ISBN: 978-1-61782-897-3

Printed from e-media with permission by:

Curran Associates, Inc.
57 Morehouse Lane
Red Hook, NY 12571

Some format issues inherent in the e-media version may also appear in this print version.

Printed by Curran Associates, Inc. (2011)

AIChE
3 Park Avenue
New York, NY 10016-5991

Phone: (203) 702-7660
Fax: (203) 775-5177

www.aiche.org

Additional copies of this publication are available from:

Curran Associates, Inc.
57 Morehouse Lane
Red Hook, NY 12571 USA
Phone: 845-758-0400
Fax: 845-758-2634
Email: curran@proceedings.com
Web: www.proceedings.com

TABLE OF CONTENTS

(6a) Upgrading Low BTU Fuels to Reduce Emissions and Enhance Performance In IC Engines

Natali Y. Ganfer and Marco J. Castaldi, Earth and Environmental Engineering, Columbia University, New York, NY

Landfills are currently the most common system used for the final disposal of solid waste. They are projected to remain the most common system in the forthcoming years. The release of Landfill Gas (LFG), mainly composed by CH4 and CO2 in an average proportion of 50%-50%, to the atmosphere constitutes an important environmental issue as methane is a GHG 20 times more powerful than CO2 and 23% of the methane emissions in the US come from landfills. On the other hand, LFG can be a free source of low BTU fuels. For these two reasons, LFG is in most cases being collected, and it can be used as IC engine fuel in order to generate electricity. However, because of the fluctuations in the composition of the gas, higher emissions of pollutants such as UHC, NOx, and CO are released. Thus the importance of lowering the emissions of these toxic gases whose expulsion is regulated by the EPA. One solution to this issue is the addition of Synthesis Gas (Syngas) -H2 and CO- to the input LFG.

Preliminary data has shown that by supplementing the input gas stream with as little as 5% Syngas, Nox emissions can be reduced up to 40%, CO up to 67%, and UHC up to 80%. Both the amounts of CO and UHC are reduced because the Syngas increases the activity of the fuel due to its higher flame speed; t he Nox formation is mitigated as a consequence of a more stable combustion at lower temperatures. More detailed data needs to be collected for different Syngas, LFG, and total input gas compositions. All the tests are being run in a Honda Engine GC160 with Maximum Power output of 2kW. Nevertheless, emission levels of 4 different engines with bigger capacity-Capstone CR200, Solar Centaur 40, Caterpillar G3516 LE-have been compared. It was observed that the results of the experiments can be extrapolated to an industrial scale as well.

See more of this Session: Thermal Processes for Waste Utilization
See more of this Group/Topical: Topical 9: Waste & Biomass Valorization

(6b) A Technoeconomic Analysis of Biodegradable Polymers Production From CO_2 and Waste Plastics

Michael Hartmann, Maoqi Feng, Thomas Andrews and Eloy Flores III, Chemical Engineering Department, Division of Chemistry and Chemical Engineering, Southwest Research Institute, San Antonio, TX

This paper evaluates the economic viability of manufacturing biodegradable polymers (mostly polycarbonates) from CO_2 and waste plastics. Waste plastics recycling are beneficial to environment and economy as well. Thermal processes have been used for feedstock recycling from waste plastics, such as polyethylene, polypropylene and polystyrene. Olefins are very useful feedstock in industry; however, it's expensive to produce olefins by using conventional hydrocarbons cracking technology. In our simulation, olefins are generated from pyrolysis of waste plastics with thermal energy supplied from a solar concentrator, thus no net CO_2 is produced in the pyrolysis. Biodegradable polymers are then produced from the polymerization of CO_2 and olefins directly in a catalytic reactor. This method provides an alternative method for CO_2 sequestration and utilization. A detailed economic analysis using VMGSim software is presented.

See more of this Session: Thermal Processes for Waste Utilization
See more of this Group/Topical: Topical 9: Waste & Biomass Valorization

(6c) Modeling of Medium Constituents for Optimized Production of Lipase From Niger Seed Oil Cake (Guizotia abyssinica) Using Artificial Neural Networks and Genetic Algorithms

Sarat Babu Imandi, Department of Biotechnology, GITAM Institute of Technology, GITAM University, Visakhapatnam, A.P., India and Hanumantha Rao Garapati, Department of Biotechnology, Anil Neerukonda Institute of Technology and Sciences, Visakhapatnam District , AP, India

A hybrid system of Artificial Neural Networks (ANN) and Genetic Algorithms (GA) was adopted to optimize the medium constituents to enhance the lipase production by *Yarrowia lipolytica* NCIM 3589 in solid state fermentation (SSF) using Niger seed oil cake as the substrate. Different microbial metabolism regulating fermentation variables (initial moisture content of the substrate, glucose, and urea) were studied to construct a '3–5–1' topology of the ANN for identifying the nonlinear relationship between fermentation variables and the enzyme yield. ANN model was subsequently optimized using GA for obtaining maximum lipase production. The usage of ANN–GA hybrid methodology resulted in a significant improvement (25 %) in the lipase yield when compared to preliminary investigations. MATLAB 7 was used for implementing the ANN–GA methodology and the corresponding program was included in the appendix.

See more of this Session: Thermal Processes for Waste Utilization
See more of this Group/Topical: Topical 9: Waste & Biomass Valorization

INFLUENCE OF ACIDS AND ALKALIS ON CELLULOSE PYROLYSIS PATHWAYS AND KINETICS

Salim M. Shaik [(a)], *Paul Nicholas Sharratt* [(a)] *and Reginald B.H. Tan* [(a,b)]

(a) *Institute of Chemical and Engineering Sciences, Singapore 627883, Singapore,*

(b) *Department of Chemical and Biomolecular Engineering, The National University of Singapore, Singapore*

1. Introduction

Cellulose pyrolysis has been the subject of many studies due to it being the largest component of most biomass sources. A phenomenological description of cellulose pyrolysis was hypothesised by Mamleev et al. [1]. The main thermal degradation pathway is via intermolecular transglycosylation reactions within the glucose monomers of cellulose [1,2]. Apart from transglycosylation, formation of liquid tar from cellulose can also occur via β-elimination. Under this mechanism, volatile acids (e.g. carboxylic acids) formed from the initial cellulose decomposition are able to attack the remaining cellulose as Bronsted acids thus catalysing heterolytic (ring-opening) reactions.

In this current work, we have studied the effects of acids (H_2SO_4, H_3PO_4 , H_3BO_3) and alkalis ($Ca(OH)_2$, NH_4OH) on the yields of levoglucosan and its derivatives (levoglucosenone, 1,4:3,6-dianhydro-α-d-glucopyranose, anhydroglucofuranose, hydroxymethyl furfural, furfural and levulinic acid). A proposed mechanism adapted from previous workers [3,4,5] and based on our current results is illustrated in Figure 1.

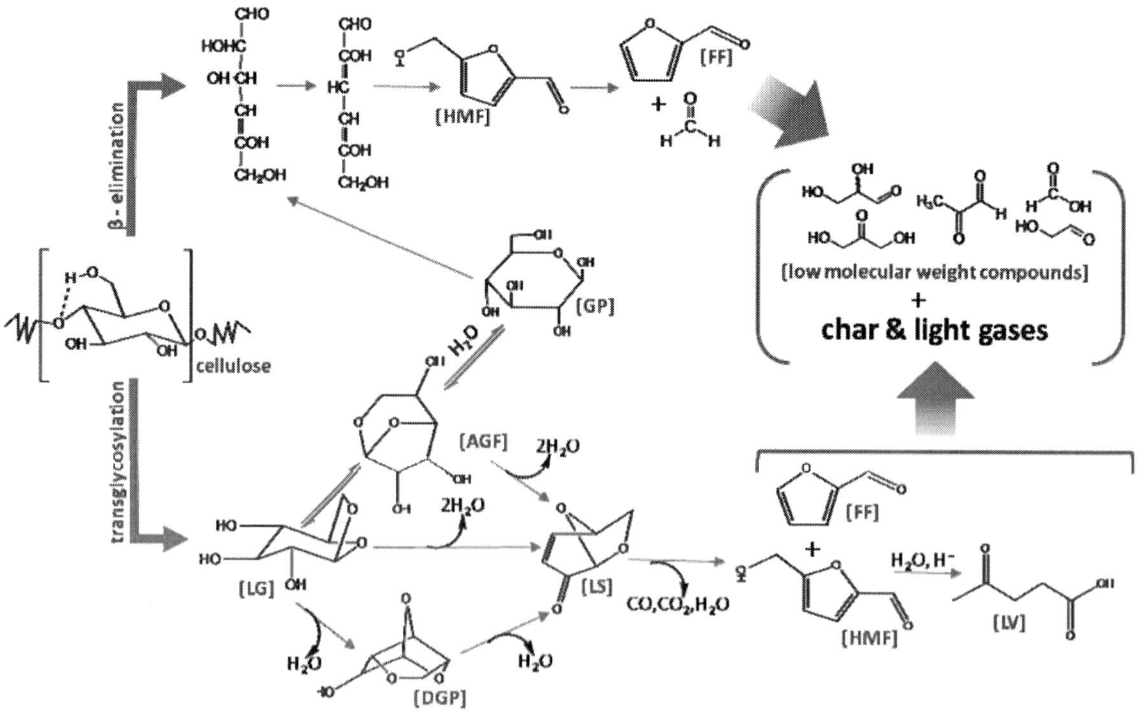

Figure 1: Schematic of possible mechanism for the conversion of cellulose to levoglucosan and its derivatives

In addition to the hypothesised role of volatile acids in pyrolytic chemistry, it has also been known that acid/alkaline content has a significant impact on the yields and product composition during direct liquefaction. Budarin and co-workers had produced bio-oils via microwave pyrolysis of wheat straw [6]. At the same time, they made use of acidic (HCl and H_2SO_4) and alkaline (NH_3) additives to effect changes in the product distribution and composition of the bio-oils derived from wheat straw pellets. The levoglucosan yields of acid and alkaline treated biomass compared with untreated biomass are shown in Table 1 below.

Table 1: Comparison of the effects of acids and alkalis on biomass pyrolysis yields.

	Budarin et al.			Hassan et al.				
additives	H_2SO_4	HCl	NH_3	H_3PO_4	H_2SO_4	NaOH	$Ca(OH)_2$	NH_4OH
levoglucosan	↓ 26%	↑ 27%	↑ 66%	↓ 54%	↓ 61%	↓ 49%	↑ 72%	↑ 41%

Budarin's results were rather mixed with no clear discernable trends between acidic and alkaline additives. A similar trend is also observed based on the results obtained by Hassan et al. [7] who studied bio-oils from pyrolysed pine wood. However, we note that the use of acids (H_2SO_4 and H_3PO_4) decreased levoglucosan yields whilst the alkalis ($Ca(OH)_2$ and NH_4OH) seem to increase levoglucosan yield. Superficially, this seems to provide some evidence for Mamleev's mechanism which suggests that acid catalysis would activate β-elimination whilst the presence of alkalis should suppress β-elimination and promote transglycosylation.

Mamleev's scheme is a novel representation of the reactions involved in cellulose pyrolysis. Although it shares some common features with other models like the Broido-Shafizadeh and Shafizadeh-Chin models, Mamleev seems to be the first to explicitly incorporate the role of acid catalysis in cellulose pyrolysis. Mamleev's scheme, indicated that there is an acid-catalysed (β-elimination) pathway with a higher activation energy ($Ea \approx 250$ kJ/mol) compared with the transglycosylation pathway ($Ea \approx 200$ kJ/mol) during cellulose pyrolysis. Based on kinetics work done by Banyasz et al. [8] and Capart et al. [9] , it was shown that the acid-catalysed, (β-elimination) pathway was favoured as temperature increases (illustrated in Figure 2).

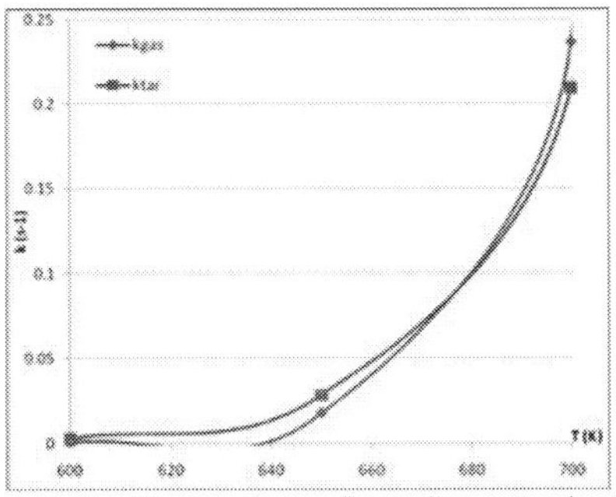

Figure 2a: Comparison of reaction rates for β-elimination (k_{gas}) and transglycosylation (k_{tar}).

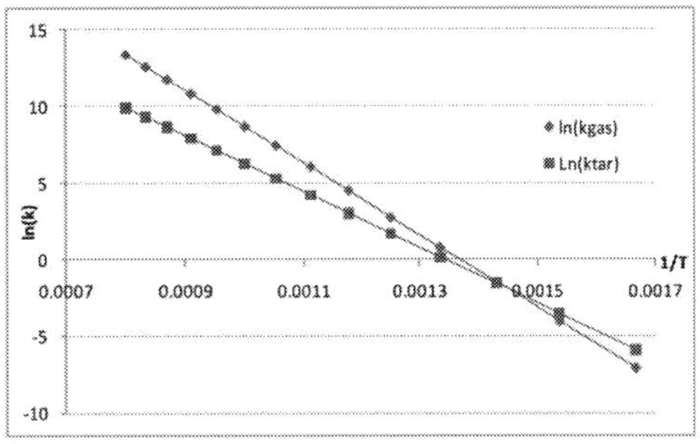

Figure 2b: Comparison of ln(k) against reciprocal temperature.

The graphs in Figure 2 also indicates that the "cross-over" temperature from transglycosylation dominated pyrolysis to ring-opening (β-elimination) dominated pyrolysis occurs at circa 670 K.

The formation of levoglucosan is expected to occur solely via the transglycosylation mechanism. It is therefore expected that the yield of levoglucosan would decrease with increasing temperature and acidity.

2. Experimental Method

The experimental setup used to study levoglucosan yield from cellulose pyrolysis consists of a 1.5 inch diameter stainless steel tubular reactor with a sample holder (capacity ca. 15cm^3) placed inside a horizontal furnace (Carbolite). The furnace is capable of heating the reactor up to 1200 °C and at a rate of 20 to 30 °C/min. Sweep and/or reacting gas is provided via two gas flow controllers which enable variations in gas compositions and reactor residence times. For the current work, nitrogen flow rates circa 650 ml/min were used and this translates to a residence time of 2 to 3 seconds.

Volatile tars and gasses that leave the reaction zone are allowed to cool immediately via three condensers (cooled to ca. 20°C, 0°C and -20°C). The condensed volatiles were analysed using a GC-MS (Agilent: 7890A GC, Column: DB-1701, 60 m x 0.25 mm I.D., 0.25 µm) for levoglucosan and anhydrosacharides content.

Cellulose samples were prepared at different acid and alkali loadings by stirring microcrystalline cellulose (Avicel from Sigma-Aldrich) with the required acid/alkali solution (0.5 to 5wt%) at ca. 25°C for 30 minutes. The "wet" cellulose samples were then filtered and dried overnight in an oven at 60°C.

3. Results and Discussion

The various components of interest namely levoglucosan **(LG)** and its derivatives (levoglucosenone **(LS)**, 1,4:3,6-dianhydro-α-D-glucopyranose **(DGP)**, 1,6-anhydro-β-D-glucofuranose **(AGF)** were identified via their mass spectrum (NIST library) and their retention times The relevant data for the standards are listed in Table 2. Quantification of the various species was done with naphthalene as the internal standard.

Table 2: Identification parameters for the main pyrolysis products

Component	LG			LS			DGP			AGF		
Retention time (min)	22.9			17.5			18.8			24.6		
Main mass fragments (m/z)	60	57	73	98	96	39	69	29	57	73	69	44

From the chromatograms, the yields of levoglucosan and its anhydrosacharide derivatives obtained from pyrolysing cellulose with varying acid/alkali loadings were calculated and compared as shown in Figures 3a to 3e.

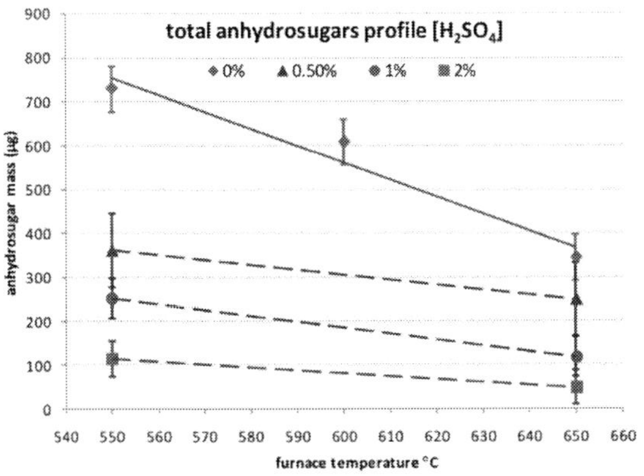

Figure 3a: Anhydrosugar yields from cellulose treated with varying concentrations of H_2SO_4 (0.5wt%, 1wt% and 2wt%)

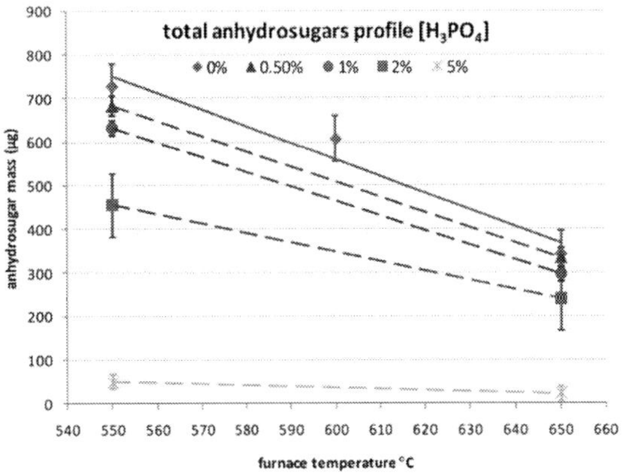

Figure 3b: Anhydrosugar yields from cellulose treated with varying concentrations of H_3PO_4 (0.5wt%, 1wt%, 2wt% and 5wt%)

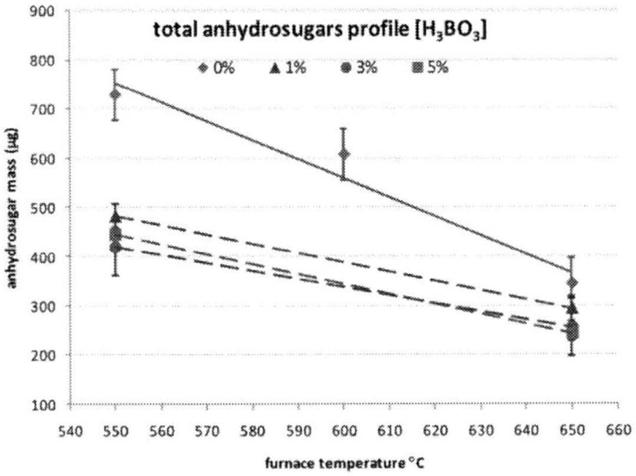

Figure 3c: Anhydrosugar yields from cellulose treated with varying concentrations of H_3BO_3 (1wt%, 3wt% and 5wt%)

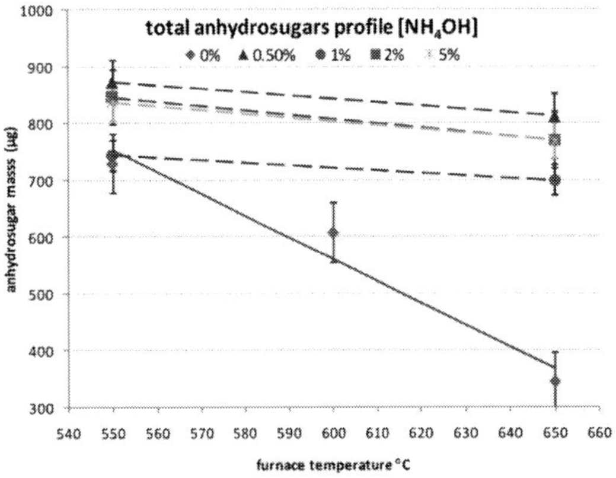

Figure 3d: Anhydrosugar yields from cellulose treated with varying concentrations of NH_4OH (0.5wt%, 1wt%, 2wt% and 5wt%)

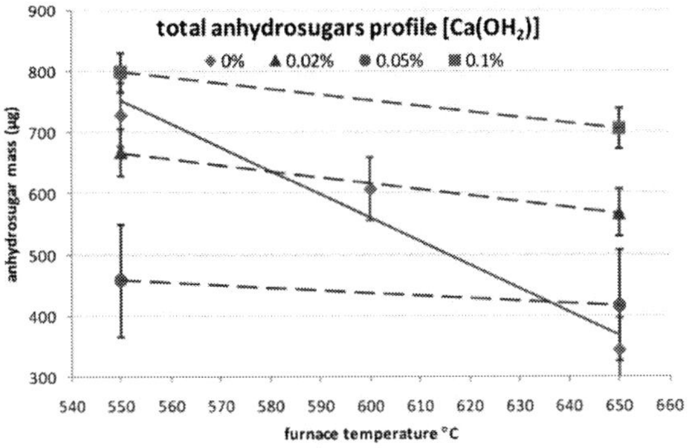

Figure 3e: Anhydrosugar yields from cellulose treated with varying concentrations of $Ca(OH)_2$ (0.02wt%, 0.05wt%, and 0.1wt%)

Looking at the graphs above, we can see that the amount of anhydrosacharides (levoglucosan, levoglucosenone, 1,4:3,6-dianhydro-α-D-glucopyranose, 1,6-anhydro-β-D-glucofuranose) decreases as the reactor temperature is increased. This would be in line with our expectations based on the relative reactions rates for transglycosylation and β-elimination. We can also see that the three acid treatments led to decreased anhydrosacharide yields with higher acid concentrations generally resulting in a corresponding reduction in anhydrosacharides.

The results obtained for cellulose pyrolysis with alkalis were slightly mixed. In the case of ammonium hydroxide, the trend was rather clear that the alkali treatment led to increased levels of anhydrosacharides being produced. However, with the calcium hydroxide treatment, higher levels of anhydrosacharides (compared to untreated cellulose) were only observed at the higher temperatures. In fact, even for ammonium hydroxide, its effects on increasing anhydrosacharides yields were also more pronounced at higher temperatures.

In order to gain further understanding from the results that we have obtained, the relative amounts of levoglucosan, levoglucosenone, 1,4:3,6-dianhydro-α-D-glucopyranose, 1,6-anhydro-β-D-glucofuranose were analysed and is represented in Figure 4.

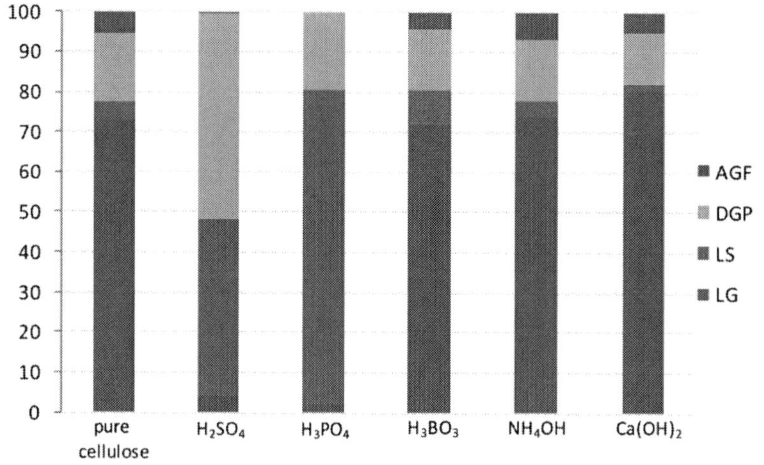

Figure 4: Comparison of the relative yields of the anhydrosacharides [levoglucosan **(LG)**, levoglucosenone **(LS)**, 1,4:3,6-dianhydro-α-D-glucopyranose **(DGP)**, 1,6-anhydro-β-D-glucofuranose **(AGF)**]obtained for the various acid and alkali-treatments.

Figure 4 clearly indicates that treatments with sulphuric and phosphoric acids resulted in higher levoglucosenone levels whereas for boric acid, ammonium hydroxide and calcium hydroxide, the relative amount of levoglucosan was found to be higher.

Based on our results thus far, we can ask the question whether a lower anhydrosacharide yield obtained in the presence of acids was due to the further breakdown of the anhydrosacharides or was it due to promotion of the acid catalysed β-elimination pathway as seen in Figure 1. If we take a closer look at the boric acid treated cellulose runs, the relative amount of levoglucosan was higher compared to the amount of levoglucosenone. This observation can be explained based on the findings of Kawamoto et al. [10] which showed that boric acid forms a stable complex with levoglucosan that prevents further transformation/breakdown. Hence in spite of this, we were still able to observe a decrease in total anhydrosacharide yield for boric acid treated cellulose. This therefore means that the cellulose probably depolymerised via the alternative β-elimination pathway.

On the other hand, we have also seen that at higher temperatures, the alkali-treated cellulose was able to maintain its anhydrosacharide yields better than untreated cellulose. This "stabilising" effect of the alkalis could be due to neutralisation of H^+ ions that are needed to catalyse the β-elimination pathway. This, in turn, allowed the transglycosylation pathway to be favoured during the pyrolysis.

4. Conclusion

Based on these experimental results, it can be seen that the presence of an acid or an alkali can have a significant effect on the yield of anhydrosacharides during pyrolysis. We have observed that acids (H_2SO_4 and H_3PO_4) decreased the total amount of anhydrosacharides and levoglucosan. This is probably due to the acids catalysing the hetrolytic reactions and promoting secondary dehydration reactions which converts levoglucosan into levoglucosenone and 1,4:3,6-dianhydro-α-d-glucopyranose. We have also seen that when the cellulose was treated with alkalis (NH_4OH and $Ca(OH)_2$), the overall yield of anhydrosacharides and levoglucosan increased accordingly.

These two main observations indicate that the β-elimination pyrolysis pathway is likely to be catalysed by the presence of acidic species at the expense of the alternative transglycosylation pathway. This is further confirmed when the presence of an alkali "inhibited" the β-elimination route in favour of transglycosylation.

References

1. **Mamleev V., Bourbigot S., Le Bras M., Yvon J.**, The facts and hypotheses relating to the phenomenological model of cellulose pyrolysis. Interdependence of the steps, *Journal of Analytical and Applied Pyrolysis*, 84, 1-17. 2009.
2. **Graham R.G., Bergougnou M.A.**, Fast pyrolysis of biomass, *Journal of Analytical and Applied Pyrolysis*, 6, 95-135. 1984.
3. **Kawamoto et. al.**, Catalytic pyrolysis of cellulose in sulfolane with some acidic catalysts, *Journal of Wood Science*, 53, 127-133. 2007.
4. **Shen D.K., Gu S.**, The mechanism for thermal decomposition of cellulose and its main products, *Bioresource Technology*, 100, 6496-6504. 2009.
5. **Lin Y.C. et. al.**, Kinetics and mechanism of cellulose pyrolysis, *Journal of Physical Chemistry C*, 113 (46), 20097–20107. 2009.
6. **Budarin et. al.**, The preparation of high grade bio-oils through the controlled, low temperature microwave activation of wheat straw, *Bioresource Technology*, 100, 6064-6068. 2009.
7. **Hassan E.M., Steele P.H.**, Characterization of fast pyrolysis bio-oils produced from pretreated pine wood, *Applied Biochemistry and Biotechnology*, 154, 182-192. 2009.
8. **Banyasz J.L., Li S., Lyons-Hart J., Shafer K.H.**, Gas evolution and the mechanism of cellulose pyrolysis, *Fuel*, 80, 1757-1763. 2001.
9. **Capart R., Khezami L., Burnham A.K.**, Assessment of various kinetic models for the pyrolysis of a microgranular cellulose, *Thermochimica Acta*, 417, 79-89. 2004.
10. **Kawamoto et. al.**, Stable complex formation with boric acid in pyrolysis of levoglucosan in acidic media, *Journal of Analytical and Applied Pyrolysis*, 82, 78-82. 2008.

Organosolv revisited – biomass valorization with new perspectives

Michael Schrems, Philipp Vejdovszky, Martin, Zeilinger, Sonja Bednarik,
Emmerich Haimer, Falk Liebner, Thomas Rosenau, Antje Potthast
Christian-Doppler-Laboratory "Advanced cellulose chemistry and analytics"
Department of Chemistry, University of Natural Resources and Life Sciences,
A-1190 Vienna, Austria
thomas.rosenau@boku.ac.at, antje.potthast@boku.ac.at

ABSTRACT
This study reinvestigated the well known Organosolv concept aiming at maximum lignin output using $scCO_2$ as co-solvent. Lignin is a still underestimated natural resource with potential application in the near future. The use of N-modified lignins as soil improvers was considered being one of the most interesting utilizations in response to progressive desertification in parts of the world. The presented process uses rye straw as a renewable source achieving up to 70% delignification with lignin purities of 80-85% and nitrogen contents of up to 4%. The nitrogen enrichment was achieved by thermal treatment of rye straw under alkaline Organosolv conditions simultaneously using aqueous NH_4OH as nitrogen source and supercritical carbon dioxide as pulping additive. The effectiveness of the used supercritical carbon dioxide could be demonstrated with a model substance study. Apocynol was used as a representative lignin model showing a significantly faster degradation in the presence of $scCO_2$ which is in accordance to a 40% lower Arrhenius activation energy.

I. INTRODUCTION
Due to fading fossil resources, valorization of renewable resources is on the rise.[1] The enormous demand for energy currently covered through petroleum refinery is the main driving force behind this process. The ongoing change to inexhaustible energy sources does not only require new technologies for biomass processing but will also lead to new industry branches for platform chemicals. Bioconversion of renewable resources by treatment with hot aqueous ethanol, known as the Organosolv process, is an old concept dating back to 1931.[2] Although the idea of separating or extracting components out of wooden materials arose already in the early 19th century, the middle of the 20th century can be considered as the time, when this concept drew attention to a larger community. With the beginning of the 1990s the interest in this research area increased remarkable focusing on the delignification of renewable materials, mainly wood. With the establishment as a promising approach for the future, new possible facets appeared in scientific databases. Organosolv pulping as a conceivable „green" alternative for established sulfur-containing processes as kraft and sulfite, generating enormous amounts of non-recyclable by-products, gained widespread interest with its peak in the nineties. In the following years the focus was directed to other possible renewable raw materials for Organosolv processes such as straw and annual plants. The latest development in this area is the use of Organosolv principles as a pretreatment step for various renewable resources to enhance the bioethanol output from polysaccharides in bacterial digestion.[3] Considerable pretreatment effects can also be obtained with simply using hot water but waive delignification to further increase the enzymatic accessibility.[4,5]

Organosolv conditions involve usually a mixture of water and an aliphatic alcohol, mainly ethanol in various mixtures. Comparable results towards yield can be achieved using methanol although higher purity hence delignification of the pulp can be achieved with ethanol.[6] Use of supercritical carbon dioxide as an additional solvent enables further improvements. Super critical carbon dioxide causes an anomalous porosity and lamellar structure within the material resulting in a higher accessibility for the ethanol water mixture.[7,8] The significantly higher solubility of carbon dioxide in ethanol in opposite to water makes the application of an alcohol water mixture even more preferable over pure hot water pretreatment.[9] Pretreatment steps with similar gases such as sulfur dioxide are also capable to enhance the achievable results but have to be considered less applicable because of their inability to reduce the carbon dioxide foot print.[10]

The everlasting issue in energy production based on renewable materials is how to make such a process profitable. Resources like straw are cheap but the energy input for the transformation to biogas is high rendering a simple conversion less feasible. As pulp resulting from Organosolv treatment of straw and other annual plants does not show sufficient quality to enter the paper and fiber industry, other main constituents need to be transformed to value-added products to sustain such a concept. Lignin, one of the main components in all naturally grown materials, shows a high potential to meet this demand although efficient usage could not be established in the past. Until now rather neglected by simply using it as fuel, various more beneficial applications are possible. The usage of Organosolv lignin as a polyphenol source is just one economically interesting alternative, provided pure isolation.

A very promising application is the use of Organosolv lignin as a source of artificial humic substances which can be applied for recultivation of degraded soils or general soil improvement. Progressive desertification in parts of the world renders this product class highly beneficial in the future. Ammonoxidation is one of the technological opportunities for converting lignin and other ligneous biomaterials into nitrogen-rich organo-mineral fertilizers that share a couple of similarities with natural humic substances. The special value of these modified lignins is the slow release of nitrogen due to different binding forms of nitrogen generated upon ammonoxidation. The fact, that substantially delignified renewable materials are in addition much easier converted to degradation products such as bioethanol and biogas, led us to the development of an organosolv-based bioconversion process maximizing the isolated amount of pure lignin. This can be achieved by applying high pressure ideally with carbon dioxide. As the subsequent ammonoxidation is usually executed in aqueous media, a significant amount of energy can be saved by avoiding the intermediate purification of the lignin. Therefore we combined these two process steps in one by performing the N-modification already during the Organosolv process replacing the water by aqueous ammonium hydroxide.

Understanding the delignification process and the effect of supercritical carbon dioxide on a molecular level is considered being crucial for the process design. First investigations towards the process occurring during organosolv pulping were done in the early 2000s by Kishimoto and Sano.[11, 12] In this work further studies were conducted to clarify the effect of super critical carbon dioxide on the delignification process. For this purpose 4-(1-Hydroxy-ethyl)-2-methoxy-phenol (Apocynol) was used as a representative lignin model substance.

II. EXPERIMENTAL

Chemicals

Chemicals were obtained from commercial sources and were of the highest purity available. Rye straw was obtained from different suppliers in Lower Austria and milled to <1mm containing an average water content of 6.3%. Apocynol was obtained via reduction of acetoguaiacone using sodiumborohydride in 52% yield.

Experimental setup

CO_2 assisted Organosolv pulping was performed in a 0.2 *l* autoclave with jacketed heater (SEPAREX, France). 12 *g* of the $scCO_2$-extraction material were mixed with 100 *ml* 50% (v/v) ethanol in water (initial phase) or aqueous ammonium hydroxide (0.1, 2.5, 5.0 and 10.0 w% NH_3, see **Table 1.**) and equilibrated for 15 minutes. After equilibration the mixture was filled into the preheated autoclave (80 °*C*) and pressurized with CO_2 to 6 *MPa*. Subsequently the heating jacket was heated to the desired temperature. After reaching the adjusted value, the system was pressurized using a hand-piston-pump (Sitec, Switzerland) and left under these conditions until the individual reaction time has elapsed. After the treatment the system was depressurized and the solid residue was washed with 400 *ml* 0.1*M* NaOH and subsequently filtrated. The residue was supposed to be enriched cellulose whereas the supernatant was supposed to bear lignin and hemicellulose. In order to separate these fractions, the supernatant was acidified with HCl to a pH lower than 3 leading to precipitation of the high molecular lignin. After filtration the supernatant as well as the two solid fractions were dried at 70 °*C* and subjected to further analysis.

General Analytics

GPC System. For GPC measurements, the system as described by Röhrling et al[5] was used: online degasser, Dionex DG-2410, Kontron 420 pump; pulse damper; autosampler, HP series 1090; column oven, refractive index (RI) detector, Shodex RI-101. DMAc/LiCl (0.9%, w/v), filtered through a 0.02 *µm* filter, was used as the eluant. The sample was injected automatically, chromatographed on two serial GPC columns, and monitored by RI detection. Molecular weight distribution and related polymer relevant parameters were calculated by software programs. The following parameters were used in the GPC measurements: flow, 1.00 *ml/min*; columns, two, PL gel mixed A LS, 20 *µm*, 7.5x300 *mm*; injection volume,: 100 *µl*; run time, 45 *min*. Data evaluation for lignin molecular weight are based on calibration wit narrow polystyrene standards (Astra 5 software). Standards used for calibration: 30 *kg/mol*, 114.2 *kg/mol*, 200 *kg/mol*, 300 *kg/mol*, 700 *kg/mol*.

Curie-point pyrolysis GC/MS was performed using a CPP-40 pyrolyser (GSG) coupled with a GC 6890 and MSD 5973 (AGILENT TECHNOLOGIES). 260 *µg* of the sample was pyrolysed at 600 °*C* (FECRALLOY™) for 10 seconds. The pyrolysate was carried by helium into the inlet (250 °*C*, split 1:20) of the gas chromatograph. Separation was achieved using a fused silica HP-5ms column (30 *m*, 0.25 *mm*, 25 *µm*) and a column flow of 1.0 $ml\ min^{-1}$. Oven program: 50 °*C* (5 *min*), 5 °$C\ min^{-1}$ to 180 °*C*, than 10 °$C\ min^{-1}$ to 280 °*C* (9 *min*). Auxiliary temperature: 250 °*C* (18 *min*), 10 °$C\ min^{-1}$ to 280 °*C* (14 *min*). Mass spectrometer: EI mode (70 *eV*); ion source 230 °*C*; quadrupole 150 °*C*, $7.6 \cdot 10^{-6}$ *Torr*. Peak assignment was accomplished using the NIST 2005 mass spectra database. Purity evaluation is based on calibration with lignin standards to which 2.5, 5.5 and 10.5 wt.-% of pulp were added. The evaluation of the polysaccharide content of the analysed samples was done by comparing the peaks of furfural and guajacole in the pyrograms. Furfural is a reliable indicator for the sugar content whereas guajacole is the main lignin fragment occurring after thermal fragmentation. Analysis of the hydroxymethyl-furfural peak as main product of hexoses showed that it had only minor effect on the overall result and was therefore not taken into account.

Elemental analysis (C, H, N) was performed at the Microanalytical Laboratory of the University of Vienna.

Analysis of ash and minerals was performed by the Institute of Waste Management at the University of Natural Resources and Life Sciences Vienna. The samples were heated up in aluminium oxide crucibles to 950 *°C* with a rate of 10 *K/min*, gas flow 150 *ml/min* (80% He/20% O_2) using 16mg material per analysis.

HPLC analysis was performed with a Dionex system (P580 pump, GINA 50 autosampler,) running with Chromeleon 4 software. Detector: UVD340U diode detector. Column: Agilent done with a Agilent Eclipse XDB-C18, Injection volume: 10 *μl*, Flow: 0.8 *ml/min*, Gradient program: 0% eluent A→ 95% eluent A (0 *min* → 15 *min*), 95% eluent A (15 *min* → 18 *min*), 0% eluent A (18 *min* → 20 *min*), Detection wavelength: 280 *nm*; Eluent A: 100% MeCN; eluent B: 100% Water.

Statistics

Statistical experiment design was done with CADEMO (Computer Aided Design of Experiments and Modeling) from BioMath GmbH using a Box-Behnken design. Design and calculation were done by Department of Applied Statistics at the University of Natural resources and Life Sciences Vienna.

Synthesis of apocynol

To a solution of acetoguaiacone (7 *g*, 42 *mmol*) in 212 *ml* 56% ethanol 3.5 *g* (93 *mmol*) of $NaBH_4$ are added in portions and afterwards stirred for 10 *h* at room temperature. Another portion of $NaBH_4$ (1 *g*, 26 *mmol*) is added and stirring is continued for additional 14 *h*. The reaction mixture is neutralized with CO_2 (dry ice!), introduced through a glass capillary with a wide diameter to prevent plugging, diluted with 200 *ml* water and extracted six times with 150 *ml* Et_2O. Combined organic phases were dried over Na_2SO_4 (about 6 *h*) and constricted to 200 *ml* under reduced pressure. The ethereal residue is transferred into a crystallization dish and recrystallized when all solvent evaporated. This step is necessary to prevent formation of dimers. The crystallized product is solved in small amounts of ethyl acetate (the unsolvable dimers are filtered off) and precipitated with hexane. The obtained white crystals are dried under reduced pressure. The remaining solution can be constricted and the yellowish crystals can be recrystallized again in the same matter to higher the yield (3,7 *g*; 52%).

^{13}C NMR (75 MHz, $CDCl_3$): 146.61 (*C*-3), 144.98 (*C*-4), 137.93 (*C*-1), 118.32 (*C*-6), 114.15 (*C*-5), 107.99 (*C*-2), 70.33 (*C*-α), 56.00 (OMe), 25.09 (*C*-β)

III. RESULTS AND DISCUSSION

The first part of this study will present investigations towards the Organosolv process. In the initial phase various chemicals e.g. acids and bases have been tested in different concentrations as additives to potentially raise the lignin output and or the purity. Alterations in the consumption of the washing solutions were conducted for the same purpose. After this phase a statistical approach was used to find optimal values for the process parameters. The second part will reveal new insights into the delignification process on a molecular level. Experiments have been done under high pressure at various temperatures. The influence of super critical carbon dioxide was investigated to understand this enhancing factor in more detail.

III.1 Development of an Organosolv process aiming at maximal lignin output

The Organosolv pulping process is known to work best with equivalent alcohol water ratios at temperatures at or slightly above 180 *°C* when high lignin output is desired. The reaction time appears to be directly proportional to the yield leading to increasing yields with increasing time.[3] With this knowledge the test parameters for the initial phase were chosen to be 180 *°C*, 16 *MPa* and 160 *min*. The rather high pressure is due to several factors. Carbon dioxide is known to change into the supercritical state about 31 *°C* and 7.3 *MPa*. More important is the requirement to keep water and also ethanol in the liquid state at this temperature. Additionally high pressures guarantee a high concentration of carbon dioxide in the ethanol-water mixture.

The first tests showed that ethanol gives only slightly better yields than methanol with comparable purity of the obtained lignin fraction. As methanol is approximately twenty percent cheaper than ethanol we continued the study with methanol as the co-solvent. Addition of acid (H_2SO_4) gave good yields for the lignin but very low yields for the cellulose fraction. As this fraction is also relevant for the economics of this concept, this additive has to be considered as unfavorable. The addition of base (NaOH) turned out to be similar unfavorable as this caused very impure lignin fractions. The effectivity of carbon dioxide was proven with two experiments under atmospheric pressure and at 160bar using nitrogen instead of $scCO_2$. The obtained yields were 38% and 13%

lower, respectively. These results indicate that the high pressure itself causes increasing yields but only carbon dioxide can raise the yield even further. Experiments with the washing solution showed that it is not possible to extract lignin with pure water. Very good results could be achieved with 1 *N* NaOH but this led to very high salt concentrations and impurities in the lignin fractions. The usage of 0.1 *N* NaOH gave just 6% lower yields but the salinity and the purity were significantly better. 1 *N* NH_4OH also showed some potential as washing solution but gave 60% lower yields rendering this alternative not very suitable. The determination of the lignin molecular weight M_w was done with GPC analysis. Due to the strong fluorescence of the samples it was not possible to determine the molecular weight with light scattering. Therefore the column was calibrated with polystyrene standards ranging from 30-700 *kg/mol*. The obtained M_w values from the analyzed lignin fractions were relatively homogenous with an average of 7.9 *kg/mol* ± 1.3. Analysis of the ash and mineral content displayed a rather low concentration within the lignin fraction with average ash content of 3.1 % ± 1.5. The next step was to find the optimal operating parameters for the described process. For this purpose a statistical approach as used to reduce the overall number of experiments and avoid systematic errors. The limits for an effective process were considered to be between 150-180 *°C* with pressures between 60-160 *bar*, reactions times between 10-160 *min* and ammonium hydroxide concentrations between 0.1 and 10%. **Table 1** gives an overview about these experiments including achieved yields and purity for the cellulose and the lignin fraction. All purity values have been determined by Curie-point pyrolysis GC/MS.

The bold black highlighted tests 8 and 22 in **Table 1** turned out to be the best pulping conditions when aiming for maximal lignin output. With a straw lignin content of about 20% both conditions lead to delignification of approximately 70% with a lignin purity of 80-85%. These results also show that there is a strong dependency towards all reaction parameters. Depending on the situation it can be chosen between a shorter but higher energy inputs or longer less intensive conditions. With average nitrogen enrichment of 4% in the lignin fraction this process can be considered as a promising concept for the production of nitrogen-rich lignins the latter being worth to become subjects of a thorough study evaluating their fertilizing and soil improving effects as well as their bio- and phytoactivity.

Table 1: *Statistical experiment design after Box-Behnken*

	T [°C]	p [*bar*]	t [*min*]	NH_4OH conc. [%]	Residual cellulosic material [%]	Polysaccharide content pulp [%]	Obtained Lignin [%]	Polysaccharide content lignin [%]
	Block points							
1	157	85	47	2.5	72.3	60.9	1.8	4.7
2	172	85	47	2.5	68.8	60.8	7.7	*n.a.*
3	157	135	47	2.5	75.0	57.5	8.0	*n.a.*
4	172	135	47	2.5	66.1	57.6	4.5	9.0
5	157	85	122	2.5	67.9	55.7	12.5	*n.a.*
6	172	85	122	2.5	59.8	56.7	4.5	12.9
7	157	135	122	2.5	70.5	53.7	2.7	*n.a.*
8	**172**	**135**	**122**	**2.5**	**65.2**	**55.8**	**14.3**	**19.7**
9	157	85	47	7.5	77.7	64.9	8.9	15.8
10	172	85	47	7.5	67.9	64.6	5.4	10.3
11	157	135	47	7.5	59.8	62.4	2.7	8.0
12	172	135	47	7.5	71.4	67.5	11.6	20.3
13	157	85	122	7.5	66.1	72.3	3.6	9.4
14	172	85	122	7.5	72.3	61.3	10.7	11.1
15	157	135	122	7.5	74.1	63.6	9.8	*n.a.*
16	172	135	122	7.5	52.7	*n.a.*	5.4	10.8
	Face points							
17	150	110	85	5.0	62.2	61.6	6.3	9.8
18	180	110	85	5.0	58.9	56.7	7.1	6.8
19	165	60	85	5.0	63.4	57.3	11.6	9.8
20	165	160	85	5.0	59.8	62.0	7.1	*n.a.*
21	165	110	10	5.0	74.1	52.3	3.6	*n.a.*
22	**165**	**110**	**160**	**5.0**	**86.6**	**68.0**	**14.3**	**14.2**
23	165	110	85	0.1	64.3	*n.a.*	8.0	*n.a.*
24	165	110	85	10.0	67.9	55.8	7.1	*n.a.*
	Center point							
25	165	110	85	5.0	67.0	67.6	8.9	*n.a.*

III.2 Investigation of the reaction kinetics through model substance studies

The aim of this study was to obtain a deeper understanding of the reactions occurring during the Organosolv pulping process in the presence of carbon dioxide. The focus was placed on the delignification process in dependence on the temperature applied and the role of supercritical carbon dioxide. For this purpose apocynol seemed to be a suitable model substance for the investigation of the delignification process. It is easily accessible through reduction of acetoguaiacone and representative due to its structure similar to lignin monomer units. [13]

The experiments were conducted in a metal reactor built for high pressure experiments which was pre-heated to 80 *°C*. 20 *mg* of apocynol (**1**) was placed in a small beaker and dissolved in 10 *ml* of a 1:1 mixture of ethanol and water. The beaker with the solution was placed in the reactor and the reaction chamber sealed afterwards. An initial CO_2 pressure of 6 *MPa* was applied to ensure just liquid phases already before reaching the adjusted temperature. The end temperature (140-180 *°C*) was reached within several minutes followed by adjusting a pressure of 16 *MPa* (t=0). At the end of the individual reaction time the heat control was turned off and the metal sheath was cooled with ice to accelerate cooling. Eventually, the orange coloured liquid obtained was analysed by HPLC.

Experiments conducted at 140 *°C*, 160 *°C* and 180 *°C* with and without supercritical carbon dioxide showed that there is an increase in degradation of the model substance with increasing temperature (cf. **Fig. 1**). But this increase is only of minor relevance compared with a much stronger degradation effect in presence of super critical carbon dioxide. As indicated in figure 1, the degradation rate at 140 *°C* with $scCO_2$ present is even faster than the reaction rate at 180 *°C* without $scCO_2$. The order of this reaction was estimated as first order. The calculation of the Arrhenius activation energy gave a significantly lower activation energy for this reaction when $scCO_2$ was applied. An E_A of 68.9 *kJ/mol* for the reaction without $scCO_2$ in comparison to 42.0 *kJ/mol* points out that the usage of $scCO_2$ can lower the activation energy of approx. 40%.

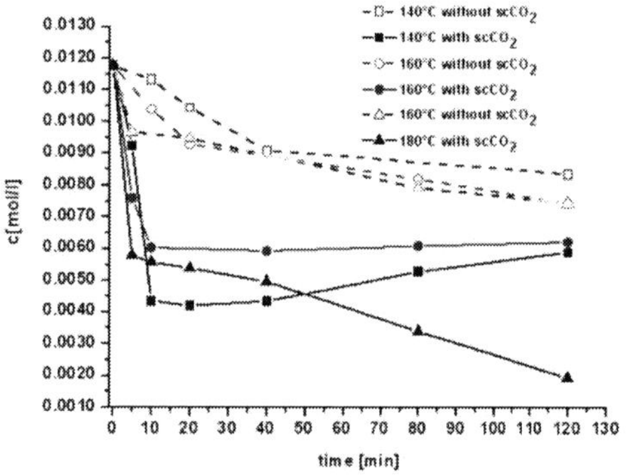

Fig. 1: Degradation of apocynol at different temperatures with and without $scCO_2$

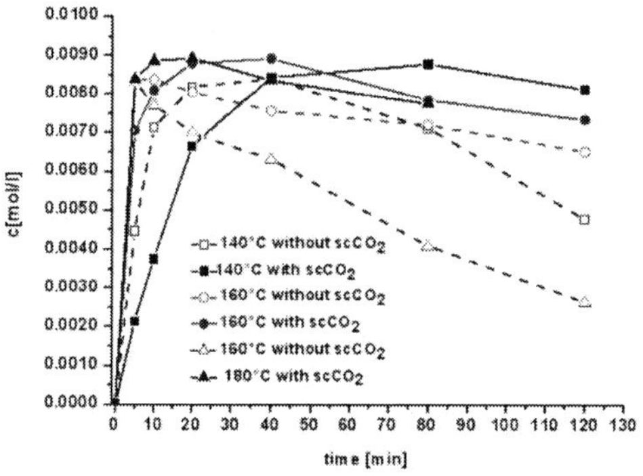

Fig. 2: Formation and transformation of the reaction intermediate 2-methoxy-4-vinylphenol

HPLC analysis of the samples revealed that the overall degradation reaction of the model substance apocynol (**1**) is a combination of two consecutive reactions. The first step is the degradation of the apocynol is the transformation into 2-methoxy-4-vinylphenol (**2**), formed via a *para*-chinone methide (*p*QM), which stabilizes into (**2**) by a [1,7]-sigmatropic proton shift.[14] The second step is the formation of 4-(1-ethoxyethyl)-2-methoxy-phenol (3) by addition of ethanol to the vinyl intermediate according to a pseudo-first-order rate law, cf. **Fig. 3**.

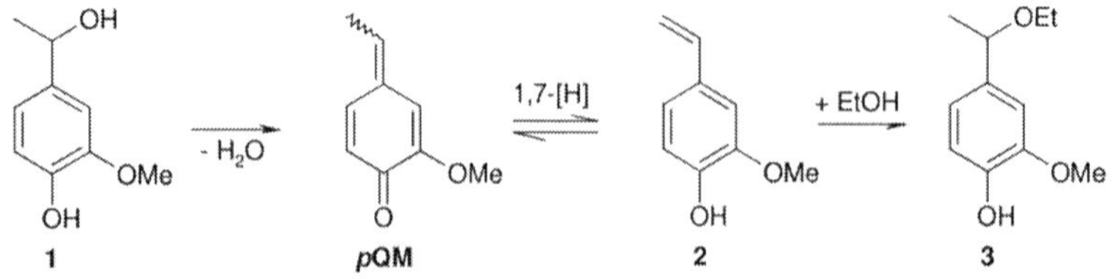

1 **pQM** **2** **3**

Fig. 3: Reaction pathway of apocynol under Organosolv pulping conditions in aqueous ethanol (50%, v/v)

The presence of $scCO_2$ slows down the subsequent transformation of the vinyl intermediate (**2**) into 4-(1-ethoxyethyl)-2-methoxy-phenol (**3**), cf. **Fig. 2**. Whereas in the absence of CO_2 the intermediate (**2**) is transformed within minutes into the final product (**3**), the presence of $scCO_2$ causes a very fast formation of the stable intermediate (**2**) but a decelerated transformation into the end product. Even the effect of temperature almost vanished. The presence of $scCO_2$ also effects the zero order formation of the end product 4-(1-ethoxyethyl)-2-methoxy-phenol (**3**). Although the final concentration is higher when $scCO_2$ is applied the initial reaction rate is definitely higher when no $scCO_2$ is present. This finding is in accordance with the calculated values for the Arrhenius activation energy. The E_A for this reaction without $scCO_2$ was determined with 4.6 *J/mol* which is about four to five times lower than the calculated value of 20.9 *J/mol* for $scCO_2$.

IV. CONCLUSIONS

The developed process allows isolating major parts of the total lignin content in rye straw. Significant increase of the lignin output can be achieved by applying high pressure during the process. Highest amounts can be obtained when $scCO_2$ is used during the reaction. This effect can be explained by the super critical state of the carbon dioxide at the investigated reaction conditions making the raw material better accessible and acting as a solvent enhancer. A statistical approach was used to find the optimal regions for all operating parameters. Temperature, pressure and time had a significant influence on the results. Additional tests were done to evaluate the contributions of various additives and auxiliaries namely acids and bases and the possible exchange of ethanol for methanol which can save up to twenty percent in the overall solvents costs.
The study showed that it is possible to achieve a delignification of approximately 70% with a sufficiently pure lignin fraction. Furthermore it was proven that nitrogen enrichment, conventionally performed subsequently, can be accomplished during the pulping process significantly reducing energy demand. A nitrogen enrichment of 4% within the lignin fraction is considered to be sufficient for the aimed purpose of soil improvement.

Kinetic studies clearly showed the influence of supercritical carbon dioxide on the reaction mechanism. Apocynol was used as a simple model for lignin. Degradation kinetics with and without super critical carbon dioxide were recorded at different temperatures between five minutes and two hours. The decay of the original model substance and the formation of degradation products were analyzed by HPLC. The use of super critical carbon dioxide during the process does not only accelerate the decay of the model substance but also lowers the activation energy distinctly, which is in accordance with current literature. It was shown that the presence of $scCO_2$ only accelerated the transformation into the intermediate product (**2**) but slowed down the formation of the end product from that intermediate. It seems quite clear that the application of super critical carbon dioxide has a positive impact on the delignification of the designed process. Tests replacing the $scCO_2$ with nitrogen within the process and the accelerated transformation of the model compound apocynol confirmed the expectations of a supportive effect. The hindrance of $scCO_2$ on the formation of the apparently end product 4-(1-ethoxyethyl)-2-methoxy-phenol can be explained with a potentially different reaction environment making the intermediate compound more stable. To get a better insight in the role of $scCO_2$ on the delignification process, a study using a dimeric compound containing a β-O-4 ether bond similar to the one used in experiments of Kishimoto and Sano is planned.[11]

V. ACKNOWLEDGEMENT

The financial support by the Christian Doppler Research Society (CD-laboratory "Advanced Cellulose Chemistry and Analytics") and Blue Globe Energy, Mannheim, Germany, is gratefully acknowledged. The authors would like to thank Prof. Dr. Ena Smidt from the Institute of Waste Management at the University of Natural Resources and Life Sciences Vienna for performing the ash and minerals content analyse and Prof. Dr. Karl Moder from the Department of Applied Statistics at the University of Natural resources and Life Sciences Vienna for the statistical experimental design.

VI. REFERENCES

1. J. H. Clark, V. Budarin, F. E. I. Deswarte, A. J. Wilson, Green chemistry and the biorefinery: a partnership for a sustainable future, *Green Chem.* **2006**, *8*, 853-860.

2. T.. Kleinert, K. Tayenthal, Ueber neuere Versuche zur Trennung von Cellulose und Inkrusten verschiedener Hölzer, *Angewandte Chemie* **1931**, *44*, 788-791.

3. X. Pan, N. Gilkes, J. Kadla, K. Pye, S. Saka, D. Gregg, K. Ehara, D. Xie, D. Lam, J. Saddler, Bioconversion of Hybrid Poplar to Ethanol and Co-Products Using an Organosolv Fractionation Process: optimization of Process Yields, *Biotechnology and Bioengineering* **2006**, *94(5)*, 851-861.

4. M. Zeng, N. S. Mosier, C-P. Huang, D. M. Sherman, M. R. Ladisch, Microscopic Examination of Changes of Plant Cell Structure in Corn Stover Due to Hot Water Pretreatment and Enzymatic Hydrolysis, *Biotechnology and Bioengineering* **2007**, *97(2)*, 265-278.

5. K. Öhgren, R. Bura, J. Saddler, G. Zacchini, Effect of hemicellulose and lignin removal on enzymatic hydrolysis of steam pretreated corn stover, *Bioresource Technology* **2007**, *98*, 2503-2510.

6. M. Oliet, J. Garcia, F. Rodriguez, M. A. Gilarranz, , Solvent effects in autocatalyzed alcohol-water pulping – Comparative study between ethanol and methanol as delignifying agents, *Chem. Eng. J.* **2002**, *87*, 157-162.

7. M. Gao, F. Xu, S. Li, X. Ji, S. Chen, D. Zhang, Effect of SC-CO_2 pretreatment in increasing rice straw biomass conversion, *Biosystems Engineering* **2010**, *106*, 470-475.

8. D. Pasquini, M. T. B. Pimeta, L. H. Ferreira, Sugar cane bagasse pulping using supercritical CO_2 associated with co-solvent 1-butanol/water, Journal of Supercritical Fluids **2005**, *34*, 125-131.

9. E. Dalmolin, E. Skovroinski, A. Biasi, M. L. Corazza, C. Dariva, O. Olivieira, Solubility of carbon dioxide in binary and ternary mixtures with ethanol and water, *Fluid Phase Equilibria* **2006**, *245*, 193-200.

10. P. Sassner, M. Galbe, G. Zacchi, Steam Pretreatment of *Salix* with and without SO_2 Impregnation for Production of Bioethanol, *Applied Biochemistry and Biotechnology* **2005**, *121-124*, 1101-1118.

11. T. Kishimoto, Y. Sano, Delignification Mechanism during High-Boiling Solvent Pulping. Part 1. Reaction of Guaiacylglycerol Ether, *Holzforschung* **2001**, *55*, 611-616.

12. T. Kishimoto, Y. Sano, Delignification Mechanism during High-Boiling Solvent Pulping. Part 2. Homolysis of Guaiacylglycerol-β-Guaiacyl Ether, *Holzforschung* **2002**, *56*, 623-631.

13. E. Adler, S. Delin, G. E. Miksche, Synthese des Isoeugenolglykol-β-(2-methoxyphenyl)-äthers, *Acta Chem. Scand.* **1966**, 20(4), 1035-1037.

14. D. Fengel, G. Wegener, Wood: chemistry, ultra structure, reactions, deGruyter, Berlin, New York, **1989**.

(14a) The Energy Challenge: Providing Sustainable Supplies to Meet Growing Demand

F. Emil Jacobs, ExxonMobil Research and Engineering Company, Annandale, NJ

As economies grow and living standards improve, the world faces a dual energy challenge: meeting the growing demand for energy while mitigating emissions and their impact on the environment. ExxonMobil believes that addressing this dual challenge requires an integrated set of solutions which includes increasing the supply of all commercially viable energy sources, reducing emissions, and improving energy efficiency in all sectors of the economy. Technology will be a key enabler of continued progress in all of these areas.

Many experts agree that the immense scale of global energy demand necessitates a continuing role for oil and gas as a substantial portion of the energy supply mix for at least the next several decades. These experts also forecast that the overall energy mix will shift toward a relatively higher level of sustainability through a combination of increased efficiency across the economy, growth in lower-carbon sources of energy, and development and deployment of technologies that help mitigate emissions.

ExxonMobil has developed a number of technologies that contribute to increased supplies of energy, reduced emissions, and higher efficiency in its operations and in consumer use of its products. The presentation will provide an overview of ongoing research and development activity in a broad range of technology areas including algae-based biofuels. It will also highlight some of company's current technologies including lithium-ion battery separator films, lightweight plastics that improve automotive efficiency, and processes that could advance carbon capture and storage.

See more of this Session: Spring Meeting Keynote Address - Sponsored by the AIChE Fuels & Petrochemicals Division
See more of this Group/Topical: Liaison Functions

COMMERCIAL SCALE TRITIUM RECOVERY FROM EXIT SIGNS AND OTHER ILLUMINATION DEVICES AT LLNL

Diane K. Spencer, J. Mark Mintz, Matthew G. Lucas and Dale M. Holck

LLNL-CONF-467413

Lawrence Livermore National Laboratory
Livermore, CA 94551
spencer2@llnl.gov, mintz1@llnl.gov, lucas3@llnl.gov

Abstract

The disposal of exit signs and similar commercially available tritium-powered illumination devices in landfills can pose potential public health and environment risks. These risks can significantly be reduced by collecting the gaseous tritium remaining in these devices. Since 2001, Lawrence Livermore National Laboratory (LLNL) and the U.S. Army have been engaged in a successful tritium recovery and recycle program for a limited number of specialized devices. This laboratory-scale work has been accomplished by hand with reasonable efficiency. However, recent requests for assistance with similar work include millions of these devices and would require much larger capabilities. Thus, LLNL has installed a commercial-scale tritium recovery system—the Tritium Grinder System (TGS)—in its Tritium Facility to complement its laboratory-scale tritium recovery capabilities. The TGS was built to efficiently process tritium-bearing illumination devices on a large scale. It can process up to 72 standard exit signs in a batch, with significantly more for smaller illumination devices (e.g., 2250 rotary phone dials per batch). The TGS liberates gaseous tritium from these various devices by mechanically grinding (or shredding) them under negative pressure in a modified industrial shredder. The liberated tritium is collected in a storage tank and periodically removed by trapping on a regenerable hydrogen getter. This paper presents the results from initial grinder operations, including the collected gas composition and unit processing rate, along with a summary of the design features, operating history, and safety controls.

Overview

Studies of active landfills show leachate with tritium considerably above background, with the tritium concentration in many above the Environmental Protection Agency drinking water MCL of 20,000 pCi/L (Ref. 1). Thus, the Nuclear Regulatory Commission (NRC) regulates tritium-containing illumination devices in general and exit signs in particular requiring reporting and handling requirements, including proper disposal of unwanted or unused signs. The requirements in most cases prevent disposal as ordinary trash. LLNL has recognized that the residual tritium in these discarded devices is valuable and has developed a method to capture it.

TGS Design and Process

The "grinder" portion of the TGS is a commercially available shredder unit from SSI Shredding Systems, Inc., modified to operate under negative pressure. A batch of devices is delivered to the grinder by a conveyor belt and is mechanically shredded. The released tritium, along with decay product helium-3 (3He), and argon (Ar) added as a sweep gas, is then pumped to a collection tank. The collected gas is processed through a commercially available getter to remove the tritium. Solid debris (e.g., exit sign housing residuals) generated through the recovery process is collected in a lined 55-gallon drum using a "bagout" process engineered to minimize outgassing and worker exposure. Upon completion of the bagout process, the content of the drum is disposed of as low-level radioactive waste.

Loading, Grinding, and Tritium Collection

Figure 1 shows the major components of the grinding and collection portion of the TGS. Illumination devices to be processed are introduced through the access door and loaded onto the conveyer belt. Small devices (e.g., telephone dials) are placed inside baskets. After the conveyor is loaded, the access door is closed and the system is pumped down. Argon purging is employed to minimize residual air and water vapor. When the system reaches a suitable negative pressure, the illumination devices and baskets are shredded, with the debris accumulating in the discharge hopper. Strategically positioned cameras allow operators to follow the progress of the baskets up the conveyor and into the grinding chamber. Evolved gas is pumped through a particulate filter to the collection tank. A series of purges with argon is then used to ensure a more complete transfer of the tritium gas from the grinder system.

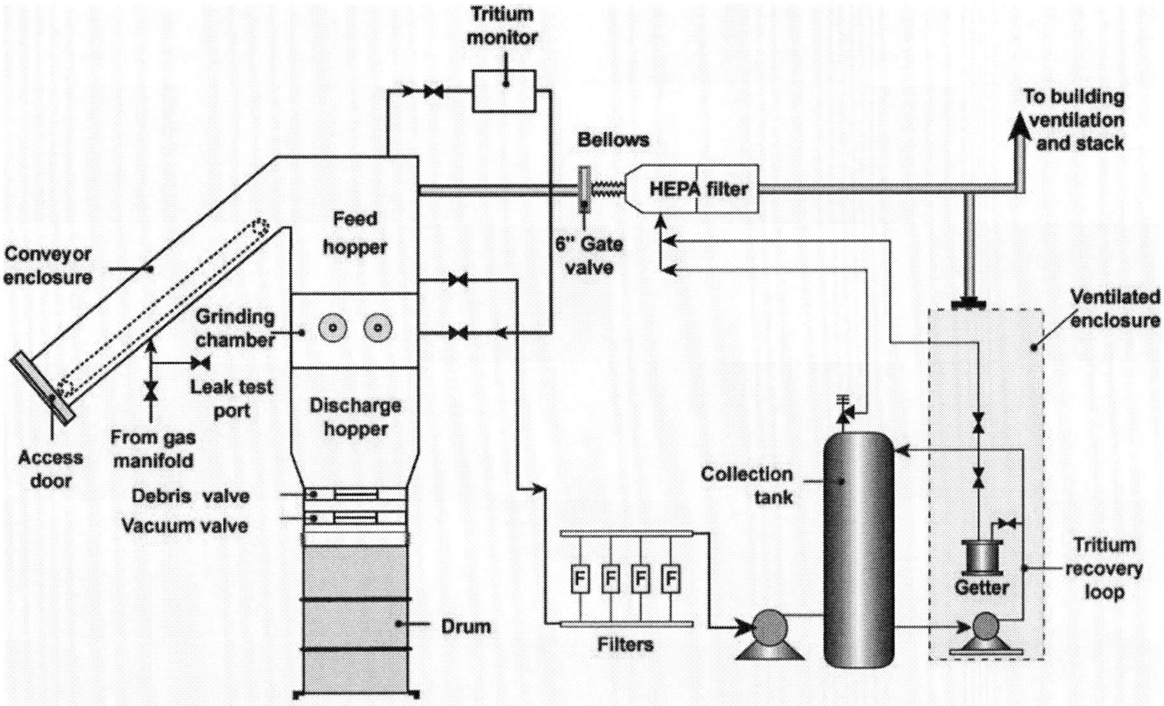

Figure 1. Major components of the TGS.

Following the final argon purge and pump out, the grinder enclosure is let up to ambient pressure, then exhausted to the building ventilation to permit tritium sampling, disposal of debris, and conveyor reloading. A dedicated tritium monitor is consulted prior to venting to verify that less than 1 Ci of tritium is released to the building ventilation when the grinder system is opened. A high efficiency particulate air (HEPA) filter protects the building ventilation from possible particulate formed during the shredding process and, in turn, the building ventilation provides airflow to protect workers from residual contamination.

The grinder is equipped with a double valve near its base. The vacuum valve is relied upon to contain tritium during processing. The debris valve above the vacuum valve supports the accumulated shredded material.

The solid debris generated during the recovery process is safely collected in a lined 55-gallon drum using a "bagout" technique similar to that developed for actinide gloveboxes. This technique allows for the waste drum liners to be sealed and replaced without external exposure. Periodically, a video camera and a small vacuum cleaner nozzle are inserted through a small slit in the liner to inspect and clean the vital vacuum valve seal. The slit is then tape sealed when the camera is removed. Following closure of the drum liner, the filled low-level waste drum is set aside for disposition, and new drum is set in its place.

Tritium Purification, Storage, and Disposition

The central feature of the tritium recovery loop is a high-capacity tritium getter (SAES CapaciTorr-B, MK5 Type). The recovery loop is used to remove tritium (and other active gases) when the pressure in the collection tank reaches 700 torr or 600 Ci, whichever is reached first. The relatively pure argon gas remaining in the collection tank after gettering is exhausted to the building ventilation, but the possibility of recycling that gas is under consideration.

Operating Experience

Grinder startup operations are conducted in two trial period phases using obsolete British Telecom rotary dial telephones used from 1964 to 1981. Phase 1 began on August 11, 2010, and concluded on September 15, 2010. During this phase, approximately 17,250 dials were processed. Phase 2 operations are ongoing, and include collecting tritium on the getter. Phase 2 is expected to be completed by April 2011. Both phases are designed to gain sufficient experience to understand the hazards associated with grinding operations and develop and implement the required controls.

Recovered Tritium

The dials processed consist of a tritium-filled, horseshoe-shaped glass tube glued onto an 84-mm diameter ABS plastic backing plate. The United Kingdom estimated that each dial contains approximately 40 millicuries (mCi) of tritium, with approximately 5 mCi as tritium gas and 35 mCi as tritium embedded in the solid matrix; the latter is not tightly bound. Data collected during Phase 1 operations showed that of the 550 Ci of tritium (HT, DT or T_2) associated with the 17,250 phone dials processed, 58 Ci were recovered. This 10.5% recovery rate compares well with the 12.5% expected from earlier measurements in the United Kingdom. The slight discrepancy may be the result of tritium labeled as methane and water known to be present in trace amounts, but not yet accounted for quantitatively.

Production Rate

Based on results to date, TGS operators can comfortably process two batches of four baskets each of low-curie illumination devices such as phone dials in an average "production" day. Operations involving higher curie items such as signs (~5 Ci each) would likely be limited to one batch a day because of a facility requirement to remove tritium from the collection tank when it approaches 600 Ci.

Collected Gas Analysis

The collection tank pressure was 320 torr at the end of Phase 1. Table 1 shows a mass spectrometric analysis of the collected gas; Table 2, the combined mole fractions of the hydrogen species. The vast majority (99%) is the purged gas Ar, followed by air constituents (O_2, N_2, and CO_2) totaling 0.86%. The presence of air is not surprising, though the O_2/N_2 fraction at 4.5, roughly the inverse of air, is interesting and could be due merely to differences in the Ar sweep efficiency.

Turning to the light isotopes, the preponderance of protium (^1H) at 93% of total hydrogen is likely the primary result of tritium exchange with environmental protium present as water vapor in the glass walls of the phone dials tritium-containment tubes (Ref. 2). Protium originating as molecular hydrogen (H_2) within the glass also may contribute to this value. The mole fraction of protium is estimated by adding the mole fractions of H_2 and half the mole fractions of hydrogen tritide (HT) and hydrogen deuteride (HD), yielding 0.0693 mole fraction %. The total hydrogen species is 0.074 mole fraction % (see Table 2).

The presence of deuterium (D2, DT, HD) can be explained as a known impurity in the original fill gas. The amount of deuterium collected is broadly consistent with the known history of these dials. For example, the approximately identical mole fractions of tritium and deuterium are reasonable given that more than 3 half lives of decay ('average' dial filled in 1972) have converted most of the original tritium fill to ^3He. Tritium half life is 12.5 years and decays to ^3He. After 38 years of decay, approximately 12% of the tritium remains. This suggests that an original deuterium fraction on this order is reasonable and well within the significant uncertainties involved in this estimate.

Table 1. Collected Gas Composition.

Compound	Mole Fraction (%)
Ar	99
O2	0.67
N2	0.15
CO2	0.033
H2	0.066
^{3}He	0.026
He	0.0085
HT	0.0035
HD	0.0032
D2	0.00078
T2	0.00045
DT	0.00036
Methanes	trace
Waters	trace
^{38}Ar	trace
Total	**100**

Table 2. Hydrogen Species Mole Fractions.

Compound	Mole Fraction (%)
Hydrogen (H2, HD/2, HT/2)	0.069
Deuterium (D2, HD/2, DT/2)	0.0026
Tritium (T2, HT/2, DT/2)	0.0024
Total Hydrogen Species	**0.074**

The observed ^3He/tritium ratio of 11 is more difficult to interpret quantitatively due to hydrogen/helium permeation rate differences through the phone dial glass containment tubes and other factors. In the absence of permeation or other differential loss mechanisms, however, a $^3He/T_2$ ratio of about 15 would be observed. This result suggests that permeation of ^3He is faster than tritium through the dial glass as would be expected (Ref. 3).

Hazards and Controls

Phase 1 grinder operations were designed primarily to gather information on the nature and type of radiological hazard the recovery process might pose. Very conservative assumptions were made about the nature and characteristics of tritium contamination that might be generated by the grinding process. In particular, the possibility that residual fine particulates incorporating with or otherwise being contaminated with tritium (i.e., Special Tritium Compound) (Ref. 4) could be a hazard to workers was taken into account. Thus, the area where grinder operations are performed is treated as an "airborne contamination area" and a very high degree of worker protection is specified. Worker protection measures include use of hooded Powered Air Personal Respirators (PAPR, 3M Model GVP-100) and tape-sealed gloves and booties, along with extensive airborne particulate and surface swipe monitoring programs.

Tritiated Particulates

Breathing zone air samplers and continuous passive air samplers ("goose neck" monitors) confirmed that airborne particulates were either extremely low in concentration or not present and therefore not a problem for grinder operations with phone dials. The breathing zone samplers have been discontinued for subsequent operations, but the continuous passive air samplers are being used in the Phase 2 trial period. Thus, the posting in the highest work area was downgraded from "Airborne Contamination Area" to "Contamination Area."

Spreadable Contamination

Consistent with past experience with manual processing of U.S. Army illumination devices, moderate levels of surface contamination (2000–5000 dpm/100 cm^2) were found in several working areas. Light decontamination (mopping) between grinding sessions has proven effective in keeping this spreadable contamination under control.

Personnel Protective Equipment

For Phase 2, use of a full-face HEPA respirator is required only when the conveyor access door is open or while performing bagout-related activities.

Path Forward

The operational trial period for TGS is expected to end April 2011. After that, exit signs and other illuminating devices as well as telephone dials will be processed routinely. Helium-3 is a valuable isotope for fusion research and other applications; the facility is exploring the feasibility of also capturing this material from obsolete illumination devices.

Acknowledgments

This work performed under the auspices of the U.S. Department of Energy by Lawrence Livermore National Laboratory under Contract DE-AC52-07NA27344.

References

1. *Lack of Tritium Exit Sign Control and Contamination of Landfill Leachate*, Association of State and Territorial Solid Waste Management Officials; ASTSWMO Radiation Focus Group, Federal Facilities Research Center, Washington, DC, July 2009.
2. *Tritium in Scottish Landfill Sites*, Report prepared for the Scottish environment Protection Agency by Galson Sciences Ltd., May 2000.
3. Shelby, J. E., Handbook of Gas Diffusion in Solids and Melts, ASM International, Materials Park, OH, April 1996.
4. DOE-HDBK-1129-2008, Tritium Handling and Safe Storage, U.S. Department of Energy, Washington, DC., December 2008.

(21b) Behavior Physics-Mechanical Properties of the Cementitious Matrix From Arum„ Vegetal Fiber, Cement, Metacaulnita and Sand

Goreti M. S. M.[1], **Felix Gamarra**[2], Ramildo D. Tolĺdo filho[1] and **Raimundo Vasconcelos**[3], (1)Civil Engineering, Federal University of Rio de Janeiro, Rio Janeiro, Brazil, (2)Department of Gama, University of Brasilia, UnB Gama College, Brasilia, Brazil, (3)Department of Technology, Federal University of Amazonas, Manaus, Brazil

Resent researches during the last three decades showed that the application of natural products have increased with more intensity. A valuable service would therefore be to find materials alternative as fibers to enhanced physic-mechanical properties of cementitious matrices. In this work was evaluated the capability of *Arumã* vegetable fiber with possibilities to reinforce cementitious matrix. The *Arumã (Ischnosiphon polyphyllus),* is found in the Amazon biome. These fibers are widely used by the tradition people of the Northern Region in handicrafts and construction. The objective this work was studied the behavior physic-mechanical of the reinforcement of cementitious matrix from *Arumã* fibers, ciment, metacaulinita and sand. We find that the values of some properties as strength, modulus of elasticity and traction in the cementitious matrix made were compatible with the values of literature and commercial standards. The results showed to highlight potential that exist in this fiber for developing of cementitious matrix

See more of this Session: Non-Thermal Valorization of Waste and Biomass
See more of this Group/Topical: Topical 9: Waste & Biomass Valorization

(21c) Influence of Process Conditions on Properties of Ash From Biomass Gasification

Naomi Klinghoffer, Earth and Environmental Engineering, Columbia University, New York, NY, Marco J. Castaldi, Department of Earth and Environmental Engineering (Henry Krumb School of Mines), Columbia University, New York, NY and Ange Nzihou, Chemical & Environmental Engineering, Ecole des Mines d'Albi-Carmaux, ALBI Cedex 09, France

Influence of process conditions on properties of ash from biomass gasification

Naomi Klinghoffer[1], Marco J. Castaldi[1], Ange Nzihou[2]

[1]Columbia University Department of Earth and Environmental Engineering

[2]Ecole des Mines d'Albi Carmaux, Department of Chemical and Environmental Engineering

Abstract

Biomass is likely to be a significant energy resource in the future. A common way to recover energy from biomass is through gasification where synthesis gas is produced; a by-product of this process is ash. This research investigates the potential to use the ash as a catalyst by understanding the properties of ash generated under different gasification conditions. Specifically, it is desired to produce a porous ash which could be used as a catalyst or as a support for more catalytically active metals.

In this work, poplar wood was gasified under CO_2, steam, and air at different reaction temperatures. Experiments were done in a fluidized bed reactor at temperatures of 750°C and 920°C and ash was recovered. BET-surface area measurements showed that ash from gasification under CO_2 had a higher surface area than ash produced from steam gasification. TGA experiments showed that gasification under CO_2 resulted in a higher mass loss compared to gasification with steam. Gasification with steam/CO_2 mixtures yielded a mass loss similar to that of steam only which could be indicative of competitive reactions between steam and CO_2.

Introduction

Gasification is a promising technology to yield a usable gaseous product from biomass or waste with high efficiency. There are three main components that are produced from gasification:

light gases (such as H_2, CO, and CO_2) that can be used for synthesis of fuels or combustion applications, tars (heavier hydrocarbons), which need to be removed as they can cause problems with downstream equipment, and ash, which is made up of minerals and metals that do not enter into the gas phase. It has been shown that the characteristics of the ash (for example, composition and porosity) can be affected by the reactive gases that are used for gasification(1). The porosity of the ash, combined with high content of metals and minerals, makes ash a good candidate to be used as a catalyst(2). It is common that ash generated from gasification of biomass or waste is either sent to a landfill or used in concrete. Use of ash in catalytic applications would increase its value as well as avoid the need for more expensive catalysts. This research investigates the potential to use the ash as a catalyst by understanding the properties of ash generated under different gasification conditions. Specifically, it is desired to produce a porous ash which could be used as a catalyst or as a support for more catalytically active metals.

Experimental

In this work poplar wood was gasified and analyzed at three different scales. Gasification was done in an in-house built fluidized bed reactor; various flow-through gases were used and results are presented here from experiments done under 10% CO_2 in N_2 and 90% H_2O in N_2. The reactor was heated up to a maximum temperature of 750°C or 920°C and held at the maximum temperature for 1 hour or 30 minutes. The top of the reactor was equipped with a porous frit therefore all ash was trapped inside the reactor and recovered after experiments. BET surface area of the ash samples was measured (Micrometrics Gemini). TGA/DSC (Setaram 111) experiments were done where the biomass samples were heated to 800°C at 20°C/min Experiments were done at different concentrations of CO_2 and H_2O in N_2. Gasification was also done in an environmental scanning electron microscope (FEI XL30) to observe the changes in physical properties of the sample during gasification. Gasification was done under air, CO_2 and H_2O and images were recorded throughout the experiments.

Results and Discussion

After fluidized bed experiments, the ash was recovered and BET-surface area was measured. Results are shown in Figure 1. When comparing surface area of ash generated under steam at 750°C, it is clear that longer gasification time leads to a higher surface area ash. However, mass recovery data shows that the mass of ash recovered is similar in both cases; for 30 minutes 5.60% of the initial mass is recovered as ash and at 1 hour 4.95% of the initial mass is recovered. TGA tests were also done with steam at 800°C (with 9%H_2O in N_2); after 30 minutes at 800°C the mass loss is 80.5% and after 1 hours mass loss is 82.7%. Therefore, both TGA tests at low H_2O concentrations and fluidized bed experiments at high concentrations of H_2O

both show that mass loss is very gradual at temperatures in the range of 750-800°C. However, the BET surface area of the ash from fluidized bed experiments showed very different surface areas based on the time spent at maximum temperature. At 1 hour the surface area of the ash sample is 621m²/g and at 30 minutes the surface area is 429m²/g. This indicates that it is possible to produce ash with significantly different properties without necessarily influencing overall production of gas/tars. This is important because if it desired to produce ash with specific properties (for example, higher porosity), one would want to do this without impacting overall gas yield.

A.

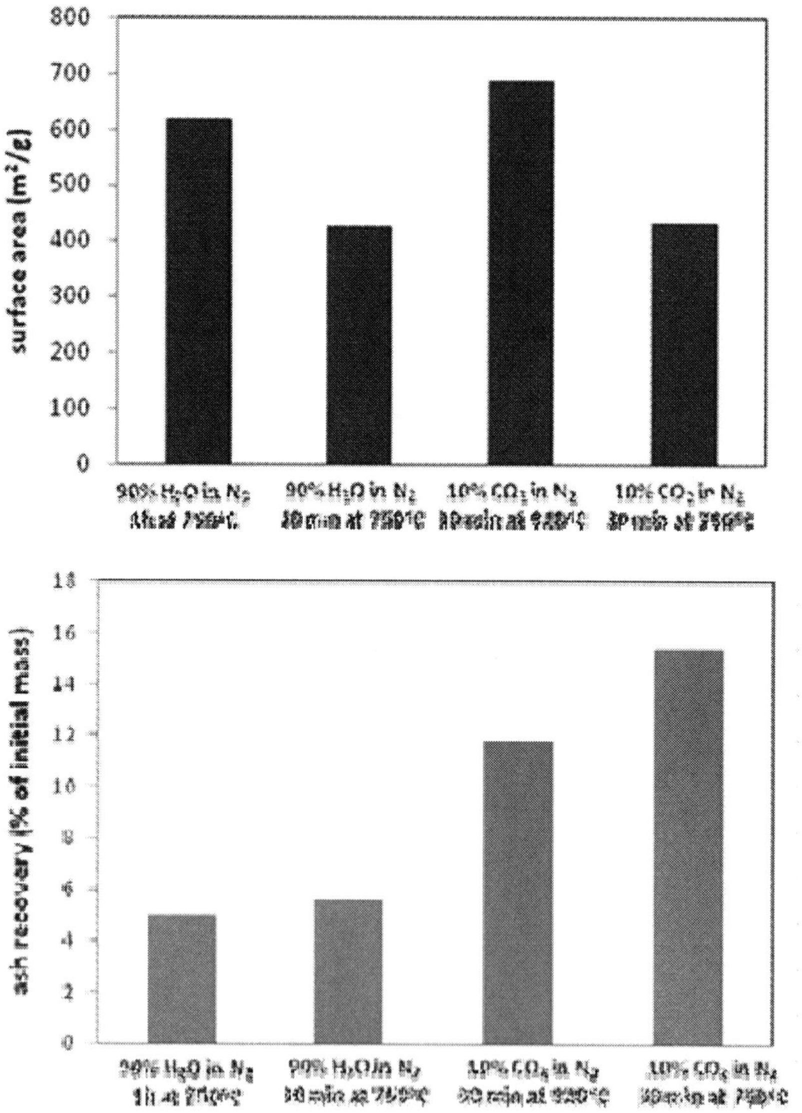

p>**Figure 1** Ash recovered from fluidized bed experiments: A. surface area B. mass recovery

Gasification was done in the fluidized bed at 750°C for 30 minutes under CO_2 (10% CO_2 in N_2) and H_2O (90% H_2O in N_2) and BET-surface area was measured. The mass recovered was 15.44% under CO_2 and 5.6% under H_2O; it is expected that the mass loss would be higher under steam since more reactant was introduced. However, the surface area of the two samples was the same. This indicates that the CO_2 produces a more porous ash per gram of biomass reacted. Therefore, if it desired to produce a more porous ash, CO_2 is likely a better candidate than H_2O.

TGA tests were done on biomass under 100% CO_2, 9% CO_2 in N_2, 9% H_2O in N_2, and $H_2O/CO_2/N_2$ mix (9%, 8%, balance), as shown in Table 1. The highest mass loss was 98.25% with 100% CO_2. It is reported in the PHYLLIS database that poplar wood has an ash content of 1.5% (dry) therefore the results with pure CO_2 demonstrate almost complete reaction. In the temperature range of 100-200°C, there is dehydration of the sample and a higher mass loss is observed for the dry samples where there is a stronger driving force for water to leave the sample. When 200°C is reached the mass loss under pure CO_2 is 7.11% while the mass loss under 9% CO_2 in N_2 is 5.85%. If dehydration is the only process taking place then one would expect equal mass loss in both cases, since both use dry gases. For example, TGA experiments were done at with air and N_2 at identical conditions (heating rate of 2°C/min to 500°C); both gases were dry. While the final mass loss was very different for the two gases (98.4% for air and 76.6% for N_2), the mass loss was similar until temperatures of 200°C. Therefore, it is possible that under CO_2 reaction begins to take place at temperatures as low as 150°C. ESEM images also showed more physical changes at low temperatures under CO_2 compared to H_2O. For example, at 400°C the pores in the biomass sample have expanded under CO_2 whereas the sample under steam has very similar physical properties to the raw biomass, as shown in Figure 2.

Table 1 Mass loss in TGA gasification experiments

Gas	Final mass loss (%)
100% CO_2	98.25
9% CO_2 in N_2	86.66
9% H_2O in N_2	81.95
9% H_2O, 8% CO_2 in N_2	81.27

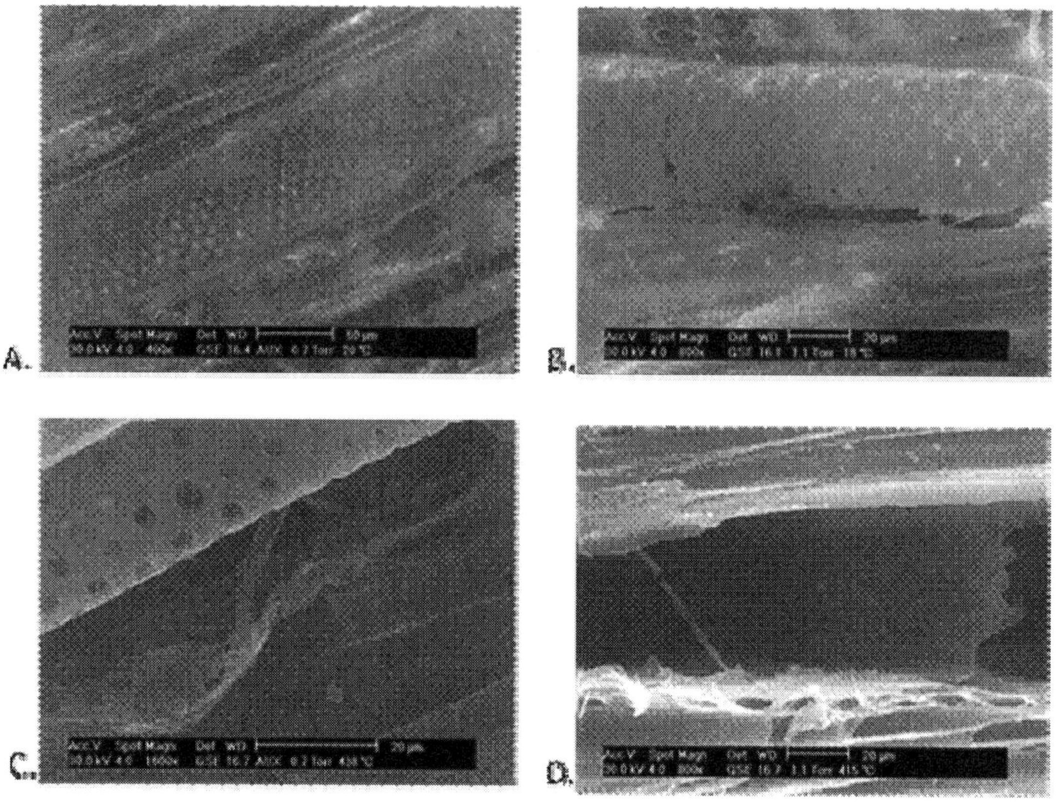

Figure 2 ESEM images A. CO_2, raw biomass B. CO_2, 430°C C. H_2O, raw biomass D. H_2O, 425°C

Mass loss was 86.66% under 9% CO_2 in N_2 and 81.95% under 9% H_2O in N_2. The higher mass loss under CO_2 could be partly due to the Boudouard reaction, shown in Equation 1. It was also interesting to note that mass loss under the mixture of CO_2 and H_2O was almost identical than mass loss under H_2O only (81.27% with CO_2 and H_2O and 81.95% with H_2O only). This could be indicative of a competitive reaction between H_2O and CO_2.

Equation 1 $CO_2 + C$ fl $2CO$ *Boudouard Reaction*

Conclusion

If the ash generated from biomass gasification is to be used in catalytic applications it is important to understand the properties of the ash as well as the ash yield that one can expect from different reaction conditions. Overall, it is desirable to produce an ash with catalytic properties without significantly impacting overall gas recovery. Gasification with CO_2 produces a more porous ash compared to gasification under steam. A higher mass loss is observed under CO_2 compared to steam and reactions under CO_2 begin at lower temperatures compared to steam.

References: (1) Butterman, H.C.; Castaldi, M. J. *Environ. Sci. Technol.* **2009**, *43*, 9030-9037

(2) El-Rub, Z. A.; Bramer, E. A.; Brem, G. *Fuel.* **2008**, 87, 2243-2252

See more of this Session: Non-Thermal Valorization of Waste and Biomass

See more of this Group/Topical: Topical 9: Waste & Biomass Valorization

(21d) Bringing Value to Waste Through Gas Fermentation

Michael A. Schultz, LanzaTech Limited, Roselle, IL

Waste gas streams are produced in many process industries, where they are either used as a fuel source, or simply are flared before being released to the atmosphere. Gas streams rich in carbon dioxide, water, carbon monoxide and/or hydrogen also can be produced through combustion or gasification of solid waste streams such as municipal solid waste, forestry residues, or agricultural residues, where the resulting gas stream is used to produce heat and/or power. In both of these cases, the main driver is to reduce the waste stream, or to convert a toxic species such as carbon monoxide to something more benign such as carbon dioxide, as well as to extract value by production of power and/or heat.

Biological fermentation of carbon monoxide rich gas streams provides an alternative route to conversion of these waste streams. The LanzaTech process produces ethanol and other high-value fuel and chemical products from renewable resources that feed on these gas streams. The carbon monoxide rich gas stream can contain any amount of hydrogen, providing an opportunity to process waste streams from a wide range of industries. The brings a significant upgrade in value to the carbon in the waste stream as compared to heat or power production, as well as an opportunity to reuse the carbon rather than release it as CO2 to the atmosphere.

See more of this Session: Non-Thermal Valorization of Waste and Biomass
See more of this Group/Topical: Topical 9: Waste & Biomass Valorization

(21e) Evaluating Different Methods Used to Manufacture Particleboards Produced Using the Blend of Plant Waste From Industrial Processes and Different Resins

Felix Gamarra, Department of Gama, University of Brasilia, UnB Gama College, Brasilia, Brazil, **Valquiria Quirino**, Forestry, Virginia Tech, Blacksburg, VA, Miriam Lima, Department of Technology, Federal University of Amazonas, Brazil, Raimundo Vasconcelos, Department of Technology, Federal University of Amazonas, Manaus and **Edison Bittencourt**, Chemical Engineering, State University of Campinas, Campinas, S„o Paulo, Brazil, Brazil

Solid wastes from industrial processes have the potential of polluting groundwater, the atmosphere, and soils. This makes its final disposal a major environmental concern. A valuable service would therefore be to find alternative uses for this waste. In this work plant waste generated by the industry was used to produce particleboards. Two types of panels were produced. The first used the blend of waste from Cumaru (*Dipteryx odorata*) and polyurethane resin. The second used the blend of waste from wood of Eucalyptus (*Eucalyptus* sp.), asparagus (*Asparagus officinalis)*, sugar-cane (*Saccharum officinarum*), and urea-formaldehyde (UF) resin. Physical-mechanical properties of the particleboards, including modulus of rupture, strength, modulus of elasticity, and perpendicular traction, were evaluated following the EN and the ASTM norms. The results obtained in this work were satisfactory. The values found were compatible to those found in the literature and in commercial standards. Thus, this experimental work demonstrated the potential use of renewable resource to produce panel for the civil industry.

Keywords: Vegetal waste, particleboard, physical-mechanical properties, resin.

See more of this Session: Non-Thermal Valorization of Waste and Biomass
See more of this Group/Topical: Topical 9: Waste & Biomass Valorization

(21f) Surface Modification of Various Substrates for Microbial Adsorption Optimization

Lee-At Sabbagh, **Diane Hildebrandt**, David Glasser and T. Matambo, Comps, University of Witwatersrand, Johannesburg, South Africa

Conversion of biomass to useful bio-products is a lengthy and often inefficient process. In our research, we have looked at the conversion of cellulose to ethanol by means of termite bacterial consortium in a fluidized bed bioreactor, where the bacteria are attached as bio-film to an activated carbon or cellulose support. Fairly high conversion rates were achieved in this reactor and the process was fairly robust and flexible. However establishing the bio-film on the activated carbon was a lengthy and fairly chaotic process in terms of the time it took the bacteria to attach to the various substrates (Activated Carbon and Cellulose) and form flocculants. The formation of such flocculants substantially increases the reaction surface and hence should optimize the production of ethanol. Bacteria carry a negative surface charge and hence for increased attraction, the surface charge of the substrate should be modified to be positive. This research, performed initially as batch processes, has shown that with the correct surface charge modifications of the substrate the electrostatic attraction forces between the surface and the bacteria are maximized. As a result of the strong electrostatic attraction forces between the two surfaces the bacteria adsorbs and attaches to the substrate quicker and creates a thicker lining on the surface than seen previously. With these results optimization of flock creation in a fluidized bed bioreactor and its effect on the production of ethanol will be further investigated.

See more of this Session: Non-Thermal Valorization of Waste and Biomass
See more of this Group/Topical: Topical 9: Waste & Biomass Valorization

(23a) Dramatic Nano-Fluidic Properties of Carbon Nanotube Membranes and Application In Chemical Processes and Water Treatement

Bruce J. Hinds, Chemical and Materials Eng., University of Kentucky, Lexington KY, KY

Carbon nanotubes have three key attributes that make them of great interest for novel membrane applications 1) atomically flat graphite surface allows for ideal fluid slip boundary conditions 2) the plasma cutting process to open CNTs inherently places functional chemistry at CNT core entrance and 3) CNT are electrically conductive allowing for electrochemical reactions and application of electric fields gradients at CNT tips. To explore the hypothesis of 'Gatekeeper' membrane selectivity, the entrances to CNT's cores were functionalized with aliphatic amines of different lengths, charged dye molecule and poly-peptides and a hindered diffusion model with a geometry of CNT tip functionalization was consistent with the experimentally observed selectivities. Changes in the flux and selectivity also support a model where charged tethered molecules at the tips are drawn into the CNT core at positive bias hindering/gating flux across the membrane. In general, the transport mechanisms through CNT membrane are a) ionic diffusion is near bulk expectation with no enhancement from CNT b) gas flow is enhanced by ~1-2 order of magnitude due to specular reflection off of flat graphitic surface c) and pressure driven flux of a variety of solvents are 4-5 ORDERS OF MAGNITUDE FASTER than conventional Newtonian flow due to atomically flat graphite planes inducing nearly ideal slip conditions. This is a nearly ideal platform to induce electro-osmotic flow with high charge density at pore entrance and a nearly frictionless surface for the propagation of plug flow. Through diazonium electrochemical modification we have successfully bound anionic surface charge to CNT tips and along CNT cores. High electro-osmotic flows of 0.05 cm/s at -300mV bias are seen with 40 fold improvements in power efficiency. Initial application demonstration in programmed drug delivery, catalytic conversion platform, bioseparations and water treatment are highlighted.

See more of this Session: University Research In Water, Bioseparation, and Pharmaceutical Field for Industry
See more of this Group/Topical: Topical 5: University Research for Industry

(23b) Fouling Resistant Membranes

Benny D. Freeman, Chemical Engineering, University of Texas at Austin, Austin, TX

An urgent need in the 21st century is access to secure, sustainable sources of fresh water. Membrane-based desalination has been recognized as an energy efficient, low environmental impact route to produce clean water for a variety of applications. Additionally, ultrafiltration and microfiltration of aqueous streams have become important unit operations in many industries, including the water purification and wastewater purification industries. This presentation focuses on 2 new classes of polymers to attack critical problems in membrane-based water purification: (1) chlorine-tolerant desalination membranes and (2) fouling-resistant membrane nanolayers based on polydopamine.

Commercially available RO membranes are prepared mainly from aromatic polyamides (PA). However, PA membranes suffer from poor resistance to continual exposure to oxidizing agents such as chlorine. Chlorine is the most widely used oxidizing biocide in water treatment because it is inexpensive and highly effective when present in water at levels of a few ppm. Disinfection of feed water to membrane desalination units is required to prevent biofilm growth on the membranes, which significantly degrades their performance. However, PA membranes cannot tolerate continuous exposure to water containing more than a few ppb of chlorine. This presentation discusses results from a systematic study of the desalination properties of a new family of highly chlorine tolerant, sulfonated polysulfones. These polymers are prepared via a novel synthesis involving direct copolymerization of non-sulfonated monomers with a disulfonated monomer. By controlling the content of disulfonated monomer in the final polymer, desalination properties, such as water permeability and salt rejection, can be varied over a wide range. A family of materials, based on random copolymers, phase-separated block copolymers, blends, and crosslinked materials may be prepared from these starting blocks to provide a new platform for desalination membranes. We are currently working with a major producer of desalination membranes to reduce these new materials to practice.

Across many platforms of membranes, fouling mitigation is a major challenge to be addressed to achieve the most energy-efficient, cost-effective membrane filtration processes. Previously, many surface modifications and functionalized polymers were reported to prevent fouling. However, most of these techniques and materials are practically difficult to implement in water purification membranes. We have discovered surface treatment methodologies that can be used to prepare high permeability polymeric membranes from all common water purification membrane classes. These surface-modified membranes have persistent tolerance to fouling by proteins and emulsified oil, two ubiquitous contaminants in a variety of wastewaters. These membranes were prepared by depositing bio-inspired, self-polymerized hydrophilic polydopamine nanolayers on their surfaces. To demonstrate scalability of this modification, the nanolayers were also applied to entire membrane modules. Upon nanolayer deposition, the membranes could be further functionalized using fouling-resistant macromolecules, such as poly(ethylene glycol), to promote improved fouling resistance and, therefore, high membrane flux. The simplicity, versatility, and broad applicability of the nanolayer deposition technique on membranes provide distinct advantages over other modification strategies to reduce fouling. A start-up company, Advanced Hydro, Inc., based in Austin, Texas has been formed to commercialize these approaches to improve membrane fouling resistance.

See more of this Session: University Research In Water, Bioseparation, and Pharmaceutical Field for Industry
See more of this Group/Topical: Topical 5: University Research for Industry

(23c) Synthesis of Reactive Nanoparticles In Membranes

D. Bhattacharyya[1], V. Smuleac[1] and Scott R. Lewis[2], (1)Chemical and Materials Engineering, University of Kentucky, Lexington, KY, (2)Chemical & Materials Engineering, University of Kentucky, Lexington, KY

Nanostructured metals have become an important class of materials in the field of catalysis, optical, electronic, magnetic and biological devices due to the unique physical and chemical properties. The nanosized metals precipitated from solutions are extremely reactive due to the high surface energy, and they usually form aggregates without the protection of their surface. Therefore, immobilization of metal nanoparticles in polymer membrane (such as, cellulose acetate, PVDF, polysulfone, chitosan, etc,) media is important from the point of view of reactivity, organic partitioning, preventing loss of nanoparticles, and reduction of surface passivation. Another major advantage of having a polymer domain is that nanoparticles (without causing agglomeration) can be directly synthesized in the matrix. This presentation will include technology developments relating to direct synthesis (including "green chemistry" approaches) of iron and bimetallic nanoparticles in microfiltration membranes for environmental applications. The project has been funded by NIEHS , US. EPA, and DOE-KRCEE.

See more of this Session: University Research In Water, Bioseparation, and Pharmaceutical Field for Industry
See more of this Group/Topical: Topical 5: University Research for Industry

(23d) The Design of Advanced Functional Membranes for Industrial Bioseparations

Scott M. Husson, Department of Chemical and Biomolecular Engineering, Clemson University, Clemson, SC

Market demand is increasing rapidly for biotherapeutics such as recombinant proteins, monoclonal antibodies, viral vaccines and plasmid DNA. With the advent of molecular biotechnology and engineered cell lines, upstream production processes have made unprecedented progress in the last decade, shifting the production burden to the downstream processing. To improve process economics and meet market demand, manufacturers will require higher productivity and higher resolution separation techniques – membrane chromatography fulfills these requirements.

In this talk, I will describe our efforts to develop adsorptive membranes for use as chromatography media. I will describe the design and performance characterization of these materials and dispel two common misperceptions by showing that (1) membrane chromatography can be a higher capacity process than resin chromatography in the purification of biologics, and (2) membrane chromatography can be a higher resolution process than resin chromatography.

Our design approach is to graft polyelectrolyte nanolayers from the pore surfaces of macroporous membranes using a controlled polymerization method. Using a number of examples, I will demonstrate that the dynamic binding capacities of proteins and DNA can be dramatically higher for well-designed macroporous membrane adsorbers than for existing commercial membrane adsorbers and resin columns. Using controlled polymerization and high polymer chain density produces adsorptive membranes with high binding capacities that are independent of flow rate, enabling high-throughput operation.

See more of this Session: University Research In Water, Bioseparation, and Pharmaceutical Field for Industry
See more of this Group/Topical: Topical 5: University Research for Industry

(23e) Microfluidic Platforms to Screen for Crystallization Conditions of Active Pharmaceutical Ingredients

Sachit Goyal[1], Michael R. Thorson[2], Yuchuan Gong[3], Geoff G.Z. Zhang[3], Charles F. Zukoski[4] and Paul J. A. Kenis[5], (1)Department of Chemical & Biomolecular Engineering, University of Illinois at Urbana-Champaign, Urbana, IL, (2)Department of Chemical & Biomolecular Engineering, University of Illinois at Urbana Champaign, Urbana, IL, (3)Solid State Sciences, Abbott Laboratories, North Chicago, IL, (4)Department of Chemical and Biomolecular Engineering, University of Illinois at Urbana Champaign, Urbana, IL, (5)Chemical & Biomolecular Engineering, University of Illinois at Urbana-Champaign, Urbana, IL

Screening for polymorphs is critical in the API development process. Identifying the available solid forms of candidate drugs (CD) as early as possible require intensive screens varying multiple parameters including counter ions, antisolvent and concentration of the CD.

Presently, highly complex robotic systems are used at the macroscale that utilize sample volumes of the order of 5 μl per condition [1,2]. Due to small quantity of the CD available in the early stages, number of conditions that can be screened is limited. Microfluidics have the potential to improve high throughput screening by utilizing reduced sample volumes, allowing improved control over mixing of fluids via free interface diffusion, and increasing the number of conditions that can be screened simultaneously [3].

Here we present a microfluidic platform enabling high throughput screening (48 wells on one platform) of antisolvents for polymorphism and crystal habit of APIs utilizing extremely small sample volumes (50nl per well). Several parameters like ratio of volume of API to antisolvent (varied from 1:5 to 5:1), solvent, and API concentration are varied to control supersaturation and are found to influence crystal habit and polymorphism. The microfluidic platform has been modified to enable in-situ Raman spectroscopy for the analysis of crystals and validation of polymorphs crystallized on-chip without need to manually harvest crystals from the chip. To support experimental results, we modeled the effect of geometry of wells on the supersaturation profiles and hence on the nucleation kinetics of the API with antisolvent. We demonstrate the capability of the chip using two model APIs – Indomethacin and Naproxen to generate different polymorphs and crystal morphologies on chip.

References

[1] P. Desrosiers, *Modern Drug Discovery*, 2004, 40.

[2] S. Talreja, D.Y. Kim, A.Y. Mirarefi, C.F. Zukoski, P.J.A. Kenis, J. Appl. Crystallography, 2005, 38, 988-995.

[3] C. Hansen, E. Skordalakes, J. Berger, S. Quake, *Proc. Nat'l Acad. Sci.*, 99, 2002, 16531-6.

See more of this Session: University Research In Water, Bioseparation, and Pharmaceutical Field for Industry
See more of this Group/Topical: Topical 5: University Research for Industry

An efficient p-chlorophenol removal prepared from cationic gemini surfactant modified montmorillonite

Xuemin Liu[1,2], Chunxu Lu[2], Cong Song[1], Songying Wang[1], Yi Wang[1]

(1.School of Chemical and Material Engineering, Jiangnan University, Wuxi Jiangsu, 214122, China; 2.School of Chemical Engineering, Nanjing University of Science and Technology, Nanjing Jiangsu, 210094, China)

Corresponding author. E-mail address: liu65xm@yahoo.com.cn

Abstract: A group of Gemini cationic surfactants containing three hydrophobic and two hydrophilic groups were used as the spacer to modify sodium montmorillonite (MMT-Na) via cation exchange to produce a *p*-chlorophenol removal. The Gemini cationic MMT was characterized with FT-IR and XRD, and its thermal stability was analyzed using thermogravimetric analysis (TG).

The modified montmorillonites performed a good removal of *p*-chlorophenol in wastewater. The *p*-chlorophenol adsorption amount of the Gemini cationic MMT is up to 39.36 mg/g, which is 7 times higher than that of MMT-Na. The modified montmorillonites also present a good thermal stability and the decomposition happened at 200-500°C mainly due to initial weight loss from residual water desorption, followed by decomposition of the surfactant and the dehydroxylation of structural water in the montmorillonite layers. The structure-function relationship of the surfactant was also investigated.

Keywords: Gemini cationic surfactant; Montmorillonite; modification; *p*-chlorophenol; adsorption

1. Introduction

Chlorophenols are widely used as pesticides and antiseptics. In addition to commercial production, chlorophenols can be produced when waste or drinking water is disinfected with chlorine. Various chlorophenols with isomers are commonly found organic pollutants in subsurface environments. Direct exposure to chlorophenols can be fatal. Long-term exposure to the chemicals may cause cancers and affect the function of the liver and the immune system. Numerous researchers have investigated methods to remove organic pollutants from aqueous phases (Lawrence et al., 1998; Wibulswas et al., 1999; Juang et al., 2004; Pernyeszi et al., 2006; Zampori et al., 2009). Adsorption of the pollutants using organo-montmorillonite has attracted much attention as one of the removal methods.

Cationic surfactants intercalated montmorillonite have gradually been the subject of extensive research, and conventional cationic surfactants were widely used as modifiers, while Gemini surfactants which contain at least two hydrophilic and two hydrophobic groups were seldom used. In fact, the montmorillonite modified by Gemini surfactants was proved to be more efficiency in removing the organic pollutes (Rosen and Li, 2001; Li and Rosen, 2000) than the one modified by conventional cationic surfactant.

In this paper, montmorillonites modified by Gemini cationic surfactants which contain three hydrophobic and two hydrophilic groups, were prepared and characterized by FTIR, XRD and TG analysis. The *p*-chlorophenol adsorption test was also carried out, expecting to be useful in removing organic pollution in wastewater.

2. Materials and methods

2.1 Materials

The montmorillonite used in this study was from ZheJiang Huate Chemical Co. Ltd (Zhejiang, China). The cation exchange capacity (CEC) was 70 mmol/100 g. A new type of Gemini cationic surfactants in Fig.1, which were synthesized according to the literature (Chen et al., 2008a; Joanna and Jan., 2006), were used as modifiers.

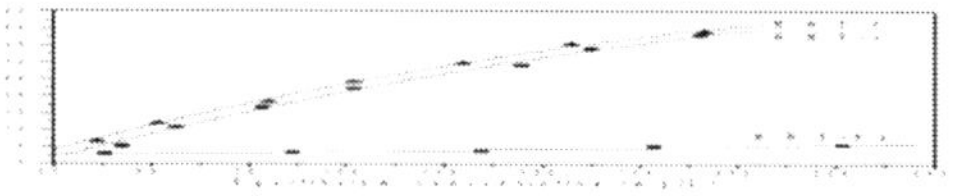

R1	n-$C_{11}H_{23}$	n-$C_{13}H_{27}$
R2	n-$C_{12}H_{25}$	n-$C_{14}H_{29}$
Surfactants	a	b

Fig.1. The structure of Gemini cationic surfactants

2.2 Preparation of modified montmorillonites

The modified montmorillonites were prepared as follows: Sodium montmorillonite power (5g) was dispersed in 100 mL of deionized water. After adjusting pH to 7.4 with hydrochloric acid, the dispersion was stirred at 80°C for 1 h. Gemini surfactants (2mmol) were dissolved in 50mL of deionized water and slowly dropped into the montmorillonite dispersion. After dropping, the dispersion was treated at 80°C for 4 h. The precipitates were separated by filtration and washed several times with deionized water until no chloride ion was detected by $AgNO_3$ solution. In this paper, MMT-a, b represent montmorillonite modified by surfactant a, b, respectively.

2.3 Characterization methods

2.3.1 Fourier transform infrared spectroscopy

The Fourier transform infrared (FTIR) spectra were obtained on FLA2000-104 FTIR spectrometer.

2.3.2 Thermogravimetric analysis

Thermogravimetric analysis (TG) were performed on a TGA Instruments (TGA /SDTA 85/e) and operated at a rate of 10.00°C/min from room temperature to 800.00°C in a high-purity flowing nitrogen atmosphere (100.0 mL/min).

2.3.3 X-ray diffraction

X-ray diffraction measurements (XRD) were performed using a D8-Advance X-ray diffractometer with CuKα radiation (λ = 1.5418 Å), operated at 40 KV and 30mA in a step scan mode. Diffraction measurements were conducted within the range of 2–10°, at a scanning speed rate of 0.02°/s.

2.4 Adsorption test

The adsorption experiments were carried out using batch equilibrium tests. 0.1g modified montmorillonite and 25 mL of prepared solution containing *p*-chlorophenol were placed in 250 mL conical flask. The adsorption experiments were carried out at

(35±0.2) °C and stirred at 180 r/min for 60 min. After separated by centrifugation, the concentration of *p*-chlorophenol was determined by Double Beam UV-VIS Spectrophotometers (λ_{max}= 280 nm).

3. Results and discussion

3.1 Characterization of modified montmorillonites

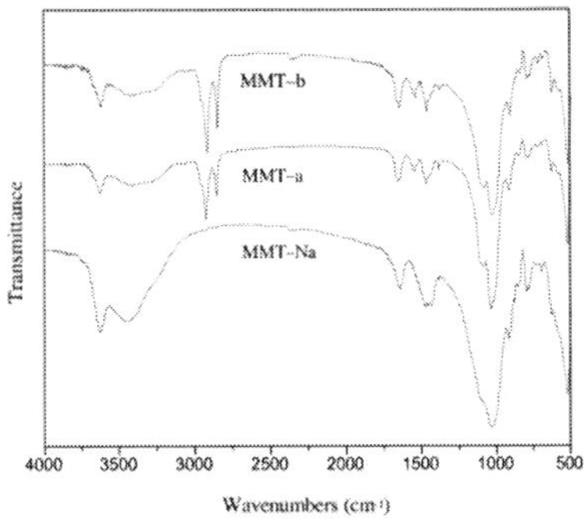

Fig.2. IR spectra of MMTs

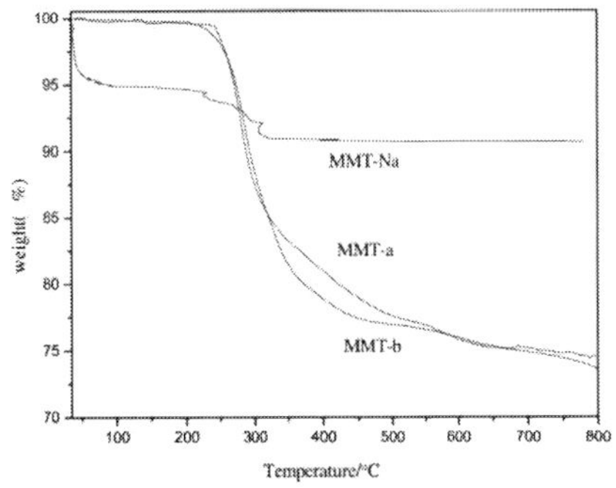

Fig.3. TG curve of MMTs

The unmodified montmorillonite is characterized by the bands of Si-O stretching and bending as well as OH bending (Zhou et al., 2009). The C-H stretching vibration of alkyl chains usually occur in the 3020-2800 cm^{-1} region (Kozak and Domka., 2004). In Fig.2, OH stretching vibration and bending vibration appears at 3640~3440 cm^{-1} region and 1639 cm^{-1} respectively. The absorption band at 1030 cm^{-1} is attributed to Si-O stretching vibration, while bands in 600~500 cm^{-1} are assigned to Si-O and Al-O bending vibration. The peaks at 2923 $cm^{-1}$and 2850 cm^{-1} are due to the asymmetrical and symmetrical C-H stretching of the methylene groups. The C-H bending vibration of the alkyl chains is observed at 1468 cm^{-1}. As is the Characteristic absorption peak of amide group, the absorption peaks that occur in 1652 cm^{-1} and 1545 cm^{-1} must be emphasized. Since there is no chloride ion to be detected, it can be proved that the surfactants have been successfully intercalated into the interlayer of montmorillonite.

Table 1. The weight loss in the thermogravimetric curves of MMTs

Samples	We.% loss at 35-200°C	We.% loss at 200-500°C	We.% loss at 750°C
MMT-Na	5.43	3.80	9.31
MMT-a	0.26	22.14	25.23
MMT-b	0.38	22.69	25.57

Thermo-analytical method provides a measure of the thermal stability of materials (Chen et al., 2008b) and has been widely used in studying the thermal decomposition of the modified montmorillonite(Xi et al., 2005a,b; Xie et al., 2001, 2002; He et al., 2005; Zhu et al., 2008). As shown by the TG curve in Fig.3, the thermal decomposition of MMT-Na occurs in two steps. The first step of weight loss is observed at 25-200°C region which is attributed to desorption of free water (water between particles and sorbed on the external surfaces of crystals) from the montmorillonite. The second step of weight loss occurs in 250-450°C regions, corresponding to the loss of water in the interlayer or to the dehydration of the hydrated cation in the interlayer. In Table 1, the organic constituent in modified montmorillonite mainly decompose at 200-500°C region, while there is little weight loss at 35-200°C region, which indicates that the free water has been replaced by Gemini cationic surfactants.

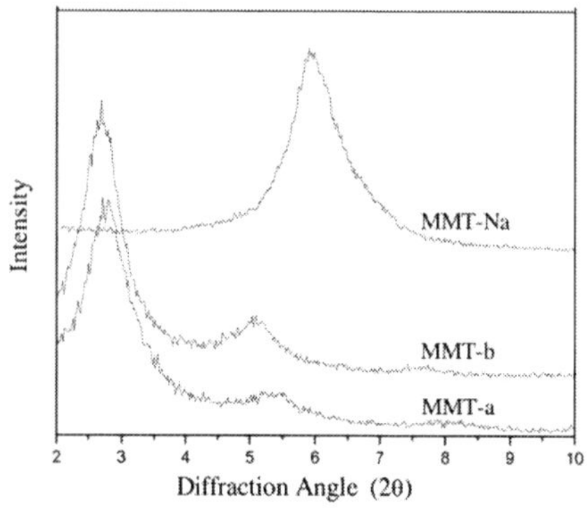

Fig.4. X-ray diffraction patterns of MMTs

According to Fig.4, the layer distance (Δd) was calculated by subtracting the sickness of the silicate layer of montmorillonite which equals 0.96 nm (Huskic et al., 2008), and listed in Table 2.

Table 2. The maximum diffraction angle and basal spacing of MMTs

Sample	Maximum diffraction angle 2θ/(°)	Basal spacing d001 (nm)	Layer distance Δd(nm)
MMT-Na	5.94	1.47	0.51
MMT-a	2.80	3.15	2.19
MMT-b	2.68	3.29	2.33

From Table 2, the basal spacing value of MMT-a and MMT-b is 3.15 nm and 3.29 nm respectively. As the carbon chain length of Gemini cationic surfactants increase, the size of the modifier augment, and the basal spacing also increases. Compared with the modifier Gemini cationic surfactant (12-4-12) reported by Zhou et al (2009), which increases the basal spacing to 2.06 nm, the modifiers used in this paper are more effective at expanding the interlayer space of the montmorillonite. According to literature (Lagaly, 1986), the alkyl chains in the modified montmorillonite interlayer might form monolayers(1.37nm), bilayers(1.77nm), pseudotrimolecular layers(2.17nm) and paraffin structure(2.77nm) as the basal spacing changes. We can conclude that these novel Gemini surfactants with more hydrophobic space groups are easier to form a paraffin structure in the montmorillonite interlayer than 12-4-12.

3.2 The adsorption isotherms of *p*-chlorophenol on montmorillonites

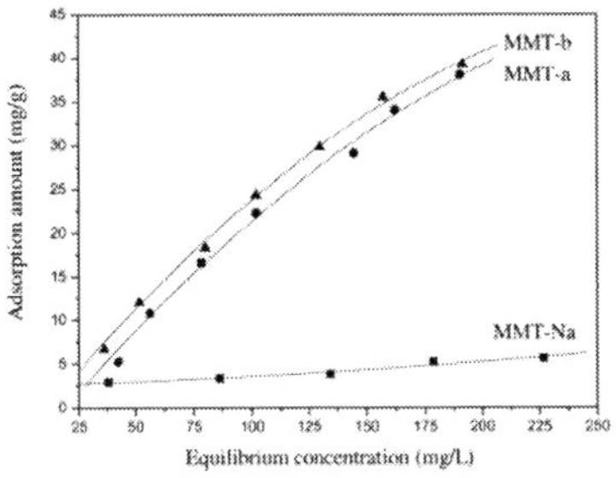

Fig.5. Isotherms of the adsorption of *p*-chlorophenol on MMTs

As shown in Fig 5, the adsorption amount of *p*-chlorophenol on MMT-Na has a linear relationship with equilibrium concentration, and the growth rate is extremely low. The modified montmorillonite did not have advantage in adsorption to *p*-chlorophenol at low equilibrium concentration, however, as the equilibrium concentration increase, the adsorption amount to *p*-chlorophenol of the modified montmorillonite increases rapidly. The adsorption amounts are up to 39.36 mg/g and 39.27 mg/g respectively at the maximum equilibrium concentration, which is seven times higher than the maximum adsorption amount of MMT-Na. It can be explained that there be more adsorption area and surface non-polarity of the montmorillonites modified by these novel Gemini surfactants. For the surface of MMT-Na is hydrophilic, it is prone to adsorb more water rather than *p*-chlorophenol, so little p-chlorophenol was adsorbed to the surface of MMT-Na, and the adsorption amount reveal slow growth as the equilibrium concentration increase. After modification, the surface of modified montmorillonite changes from hydrophilic to hydrophobic, the surface energy reduced. With a paraffin structure, the interlayer of the modified montmorillonite exhibited good compatibility with *p*-chlorophenol, adsorption and partition together affect the adsorption of *p*-chlorophenol on the modified montmorillonite, leading to high adsorption to *p*-chlorophenol.

4. Summary

FTIR analysis proves that the Gemini cationic surfactants have been successfully intercalated into the montmorillonite. The TG curve shows that there is little mass loss at 35-200°C regions, while the organic constitutes mainly decomposed in 200-500°C region. The XRD results show that the surfactants used in this paper can expand the basal spacing of the montmorillonite effectively and have more advantages than the other Gemini cationic surfactants. The two intercalators form a paraffin structure in the montmorillonite interlayer. The modified montmorillonite revealed a better removal of p-chlorophenol than the unmodified montmorillonite in adsorption test, which showed a promising application on the removal of the organic pollutes in wastewater.

References

Chen, F.S., Liu, X.M., Hu, Z.R., Zhao, C.X., 2008a. Synthesis and properties of cationic Gemini surfactants with amide group. *China Surfactant Detergent & Cosmetics*. 38 (5) 285-288.

Chen, D., Zhu, J.X., Yuan, P., Yang, S.J., Chen, T.H., He, H.P., 2008b. Preparation and characterization of anion-cation surfactants modified montmorillonite. *J. Therm. Anal. Cal.* 94, 841-848.

He, H.P., Ding, Z., Zhu, J.X., Yuan, P., Xi, Y.F, Yang, D., Frost, R.L., 2005. Thermal characterization of surfactant-modified montmorillonites. *Clays and Clay Minerals*. 53,287-293.

Huskic, M., Brnardic, I., Zigon, M., Ivankovic, M., 2008. Modification of montmorillonite by quaternary polyesters. *Journal of Non-crystalline Solids*. 354, 3326-3331

Joanna, W., Jan, C., 2006, Preparation, Surface-Active and Antielectrostatic Properties of Multiple Quaternary Ammonium Salts. *Journal of Surfactants and Detergents*. 9 (3), 221-226.

Juang, R-S., Lin, S-H., Tsao, K-H., 2004. Sorption of phenols from water in column systems using surfactant-modified montmorillonite. *Journal of Colloid and Interface Science*. 269, 46-52.

Kozak, M., Domka, L., 2004. Adsorption of the quaternary ammonium salts on montmorillonite. *Journal of Physics and Chemistry of solids*. 65, 441-445.

Lagaly, G., 1986. Interaction of alkylamines with different types of layered compounds. *Solid State Ionics*. 22, 43-51.

Lawrence, M.A.M., Kukkadapu, R.K., Boyd, S.A., 1998. Adsorption of phenol and chlorinated phenols from aqueous solution by tetramethylammonium- and etramethylphosphonium-exchanged montmorillonite. *Applied Clay Science*. 13, 13-20.

Li, F., Rosen, M.J., 2000. Adsorption of Gemini and Conventional Cationic Surfactants onto Montmorillonite and the Removal of Some Pollutants by the Clay. *Journal of Colloid and Interface Science*. 224, 265-271.

Pernyeszi, T., Kasteel, R., Witthuhn, B., Klahre, P., Vereecken, H., Klumpp, E., 2006. Organoclays for soil remediation: Adsorption of 2, 4-dichlorophenol on organoclay/aquifer material mixtures studied under static and flow conditions. *Applied Clay Science*. 32, 179-189.

Rosen, M.J., Li, F., 2001. The Adsorption of Gemini and Conventional Surfactants onto Some Soil Solids and the Removal of 2-Naphthol by the Soil Surfaces. *Journal of Colloid and Interface Science*. 234, 418-424.

Wibulswas, R., White, D.A., RAUTIU, R., 1999. Adsorption of phenolic compounds from water by surfactant-modified pillared clays. *Trans IChemE*. 77, 88-92.

Xie, W., Gao, Z.M., Pan, W.P., Hunter, D., Singh, A., Vaia, R., 2001. Thermal Degradation Chemistry of Alkyl Quaternary Ammonium Montmorillonite. *Chem.Mater.* 13, 2979-2990.

Xie, W., Xie, R.C., Pan, W.P., Hunter, D., Bryan,Koene., Tan, L.S., Vaia, R., 2002. Thermal stability of Quaternary phosphonium modified Montmorillonite. *Chem.Mater.* 14, 4837-4845.

Xi, Y., Martens, W., He, H., Frost, R.L., 2005a. Thermogravimetric analysis of organoclays intercalated with the surfactant octadecyltrimethylammonium bromide. *J. Therm. Anal. Cal.*

81,91-97.
Xi, Y.F., Frost, R.L., He, H.P., Kloprogge, T., Bostrom, T., 2005b. Modification of Wyoming Montmorillonite Surfaces Using a Cationic Surfactant. *Langmuir.* 21, 8675-8680.
Zampori, L., Gallo Stampino, P., Dotelli, G., 2009. Adsorption of nitrobenzene and orthochlorophenol on dimethyl ditallowyl montmorillonite: A microstructural and thermodynamic study. *Applied Clay Science.* 42, 605-610.
Zhou, L.M., Chen, H., Jiang, X.H., Lu, F., Zhou, Y.F., Yin, W.M., Ji, X.Y., 2009. Modification of montmorillonite surfaces using a novel class of cationic gemini surfactants. *Journal of Colloid and Interface Science.* 332, 16-21.
Zhu, R.L., Zhu, L.Z., Zhu, J.X., Xu, L.H., 2008. Structure of cetyltrimethylammonium intercalated hydrobiotite. *Applied Clay Science.* 42, 224-231.

(40a) Energy and Greenhouse Gas Emissions of Pyrolysis Pathways

Jeongwoo Han, Amgad Elgowainy, Ignasi Palou Rivera, Michael Wang and Jennifer M. Dunn, Energy Systems Division, Argonne National Laboratory, Argonne, IL

Pyrolysis of biomass to produce liquid fuels can play a significant role toward achieving the goal of 36 billion gallons per year of biofuels by 2022 as set by the Energy Independence and Security Act (EISA) of 2007. Argonne National Laboratory conducted life cycle analysis of various pyrolysis pathways by expanding and employing the Greenhouse Gases, Regulated Emissions, and Energy Use in Transportation Model (GREET). The fuel cycle energy use and greenhouse gases (GHG) emissions from the pyrolysis pathways were compared to the baseline gasoline pathway as well as alternative biofuel pathways such as corn and cellulosic ethanol, cellulosic biomass to Fischer Tropsch Diesel (FTD), and soybean biodiesel, renewable diesel and renewable gasoline. The life cycle results are significantly impacted by the source of hydrogen production for hydrotreatment and hydrocracking processes, and the handling of the biochar and other coproducts of the pyrolysis pathway.

See more of this Session: Advances In Alternative BioEnergy
See more of this Group/Topical: Topical 9: Waste & Biomass Valorization

(40d) Efficient Fermentation of Glucose and Xylose Mixtures Using Immobilized Saccharomyces Cerevisiae on Pine Wood Chips and Free Pichia Stipitis

Ravi Dhabhai, Chemical Engineering, Malaviya National Institute of Technology Jaipur, Jaipur, India, **S. P. Chaurasia**, Chemical Engineering, Malaviya National Institute of Technology, Jaipur, India and Ajay K. Dalai, Chemical Engineering, University of Saskatchewan, Saskatoon, SK, Canada

In this work, the ability of pine wood chips (PWC) as yeast support has been evaluated. *Saccharomyces cerevisiae* was immobilized on this biological support and mixed sugar fermentation was carried out with free culture of *Pichia stipitis* and immobilized *S. cerevisiae*. Pine wood was used in two forms; untreated (UPW) and treated (TPW). Fermentation experiments with free culture of both the yeasts were also carried out. The immobilization was carried out by co-culturing 48 h grown *S. cerevisiae* culture with PWC for 24 h with gentle shaking. The immobilization was confirmed by optical microscope and by the efficiency to ferment glucose quickly as compared to free culture. Initial experiments were carried out with 50 g/l mixture of glucose and xylose (1.5:1.0 ratio of glucose to xylose).

The efficiency of support for fermentation was evident by the fact that 63-70% of sugars were consumed in only 12 h and 84-93% fermentation efficiency was observed with almost complete utilization of both the sugars in 44 h. Doubling the sugar concentration (100 g/l) had little effects on fermentation efficiency but time required for complete sugar fermentation was increased to 78 h from 44 h. TPW worked out as a better support than UPW as the surface area of the former increased after the treatment to 1.5 times.

See more of this Session: Advances In Alternative BioEnergy
See more of this Group/Topical: Topical 9: Waste & Biomass Valorization

(40e) A Review of Biodiesel Production Via Reactive Distillation

Iman Noshadi[1], Nor Aishah Saidina Amin[1], Richard Parnas[2] and Seyed Masoom Khazraee[3], (1) Chemical Engineering, University Technology Malaysia, Johour, Malaysia, (2)Institute of Material Science, University of Connecticut, Storrs-Mansfield, CT, (3)Fars Engineering Research Center, Shiraz, Iran, Shiraz, Iran

Abstract— there are two main problems with continued use of fossil fuels, which meat about 80% of the world energy demand today. The first problem is that they are limited in amount and sooner or later will be depleted. The second problem is that fossil fuels are causing serious environmental problems. Since biodiesel is sustainable and combines high performance with environmental benefits it has become increasingly attractive as an alternative fuel. The conventional method for the production of biodiesel has several disadvantages: high alcohol demand due to shift equilibrium toward fatty esters, necessity of catalyst neutralization hence causing salt waste stream, high separation costs of fatty esters products from the reaction mixture, high production costs caused by complex processes including one or two reactors and several separation units. The reactive distillation (RD) process for the production of biodiesel overcomes these disadvantages. This paper attempts to review production of biodiesel by reactive distillation. The effect of various parameters feed ratio, type of applied catalyst, feed temperature, reboiler duty, type of column have also been summarized in this paper.

See more of this Session: Advances In Alternative BioEnergy
See more of this Group/Topical: Topical 9: Waste & Biomass Valorization

(62a) Setting Biorefinery Manufacturing Fundamentals to Produce a Portfolio of Commodity and Fine Chemicals

Alain A. Vertés[1], Masayuki Inui[1] and Hideaki Yukawa[2], (1)Molecular Microbiology and Biotechnology Group, Research Institute of Innovative Technology for the Earth (RITE), Kizugawa, Japan, (2)Molecular Microbiology and Biotechnology Group, RITE, Kizugawa, Japan

The chemurgy concept, that is, manufacturing industrial commodity and chemical products from organic raw materials such as agricultural biomass and lignocellulosic waste, predates the petrochemical industry. However, petrochemicals prevailed over fermentation chemicals given critical industrialization factors, comprising a low cost of the petroleum feedstock and a high process economic efficiency, as exemplified by the full utilization of products and by-products from petrochemical plants for the production of gasoline. These economies of scale and scope have played an important role in the booming industrial development at the beginning of the XXth century and were realized on a very large scale in World War II and the following decades[1].

These extremely efficient cost structures that have been derived in the petrochemical industry over a century are now both a blessing and a curse. The blessing is that these economies of scale and scope have made possible the manufacturing of cheap fuels and materials. The resulting abundant supply of chemicals for industry and transportation has enabled modern life and thus now appear to be indispensable. The curse is that efforts to displace and replace the petroleum industry by the chemurgy industry for alleviating global warming and fossil fuel supply threats meet a very high economic and resistance barrier[2]. This has resulted over the years in a detrimental delay in renewable chemistry technology investment, development, and deployment[3].

The biorefinery concept aims at recreating for agricultural feedstock cost structures that are similarly enabling as the ones at play in the petrochemical industry[4, 5]. This change is of course fundamental and will have far reaching consequences, as it integrates the agricultural value chain into the petrochemical value chain, thus colliding the transportation, polymer, and chemical value chains with other established value chains and notably of the food and feed industry[3]. A key difference nevertheless remains in that the higher oxygen content of biomass-based compounds makes possible to produce totally novel materials and chemicals with novels properties[6]. An enabling factor of the biorefinery vision is of course the availability of versatile, cost-effective, and robust industrial processes to manufacture an array of chemicals to serve various sectors including the fuel, chemical, industrial polymer, feed, food, cosmetic, and pharmaceutical industries. Notably, it is important for the biorefinery plant manager to be able to produce in addition to fuels, such as ethanol or isobutanol, higher value compounds for the chemical industry, such as ethyl-acetate or isobutanol.

We report here on a growth-arrested biotechnological process to manufacture ethanol, organic acids, and amino acids from mixed sugars, based on *Corynebacterium glutamicum*, an organism that has a long history of use to produce amino acids as feed and food supplements or cosmetic ingredients. The intrinsic design of the process is to decouple in a separate fermentor and a reactor biological catalyst-production and product-production phases[7] with cell recycling and using very high cell concentrations; it builds upon the intrinsic properties of this organism. *C. glutamicum* is a facultative anaerobic bacterium that grows in the absence of oxygen only in media containing a terminal electron acceptor such as nitrate[8]. *C. glutamicum* does not sporulate but remains metabolically active when placed under growth-arrested conditions[9, 10] and is not inhibited by common fermentation inhibitors present in typical lignocellulosic hydrolysates[11, 12]; moreover, it can catabolize a range of sugars in parallel with glucose[7, 13, 14]. A complete toolbox of molecular tools enables one to modulate at will its plastic genome and general metabolism[10, 15]. The RITE process presents the entire set of fundamental attributes necessary for reaching cost-effectiveness and enabling a rich product portfolio: simplicity, industrial robustness[16], raw material versatility (glucose, fructose, glucuronic acid, glucosamine, maltose, mannose, α-methyl-glucoside, ribose, sucrose, trehalose, arbutin, salicin, cellobiose, arabinose, xylose)[13, 14, 17, 18], product versatility (ethanol, organic acids, amino acids)[9, 16, 19, 20], reduced utilities requirement (reduced fermentation heat[12], no oxygenation requirement[9]), high conversion rates of carbohydrate feedstocks-to-product given cell recycling and the absence of growth requirements (e.g., 110 gl^{-1} or 40 $gl^{-1}h^{-1}$ D-lactate)[21], possibility to conduct continuous fermentations for long periods of time, dramatically reduced risks of contamination in the reaction vessel given the use of very high cellular concentrations, secreted product, no expensive reagent, portfolio of processes that enable the practitioner to manage the biorefinery plant in swiftly interchangeable processes or to capture synergies by implementing multiplex productions[22].

The stage is now set for this core technology to be translated into a biorefinery demonstration plant, as a first step towards full-scale commercialization.

References

1. Benninga, H. A history of lactic acid making. (Kluwer Academic Publishers, Dordrecht, The Netherlands; 1990).
2. Vertès, A.A., Inui, M. & Yukawa, H. Implementing biofuels on a global scale. Nat Biotechnol 24, 761-764 (2006).
3. Vertès, A.A., Qureshi, N., Blaschek, H.P. & Yukawa, H. (eds.) Biomass to biofuels: strategies for global industries. (Wiley, Chichester, UK; 2010).
4. Cherubini, F. The biorefinery concept: using biomass instead of oil for producing energy and chemicals. Energy Conversion Manag 51, 1412-1421 (2010).
5. Octave, S. & Thomas, D. Biorefinery: toward an industrial metabolism. Biochimie 91, 659-664 (2009).
6. Marquardt, W. et al. The biorenewables opportunity: toward next generation process and product systems. AIChE J 56, 2228-2235 (2010).
7. Vertès, A.A., Inui, M. & Yukawa, H. in 14th European Biomass Conference 1068-1071, Paris,

France; 2005).
8. Nishimura, T., Vertès, A.A., Shinoda, Y., Inui, M. & Yukawa, H. Anaerobic growth of Corynebacterium glutamicum using nitrate as a terminal electron acceptor. Appl Microbiol Biotechnol 75, 889-897 (2007).
9. Inui, M., Kawaguchi, H., Murakami, S., Vertès, A.A. & Yukawa, H. Metabolic engineering of Corynebacterium glutamicum for fuel ethanol production under oxygen-deprivation conditions. J Mol Microbiol Biotechnol 8, 243-254 (2004).
10. Yukawa, H., Inui, M. & Vertès, A.A. in Amino acid biosynthesis, Vol. 5. (ed. V.F. Wendisch) (Springer Verlag, Berlin, Germany; 2006).
11. Sakai, S. et al. Effect of lignocellulose-derived inhibitors on growth of and ethanol production by growth-arrested Corynebacterium glutamicum R. Appl Environ Microbiol 73, 2349-2353 (2007).
12. Inui, M. et al. in 28th Symposium on Biotechnology for Fuels and Chemicals, Nashville, TN, USA; 2006).
13. Kawaguchi, H., Sasaki, M., Vertes, A.A., Inui, M. & Yukawa, H. Engineering of an L-arabinose metabolic pathway in Corynebacterium glutamicum. Appl Microbiol Biotechnol 77, 1053-1062 (2008).
14. Sasaki, M., Jojima, T., Inui, M. & Yukawa, H. Simultaneous utilization of D-cellobiose, D-glucose, and D-xylose by recombinant Corynebacterium glutamicum under oxygen-deprived conditions. Appl Microbiol Biotechnol 81, 691-699 (2008).
15. Vertès, A.A., Inui, M. & Yukawa, H. Manipulating corynebacteria, from individual genes to chromosomes. Appl Environ Microbiol 71, 7633-7642 (2005).
16. Vertès, A.A., Inui, M. & Yukawa, H. Technological options for biolofical fuel ethanol. J Mol Microbiol Biotechnol 15, 16-30 (2008).
17. Yukawa, H. et al. Comparative analysis of the Corynebacterium glutamicum group and complete genome sequence of strain R. Microbiology 153, 1042-1058 (2007).
18. Kawaguchi, H., Vertès, A.A., Okino, S., Inui, M. & Yukawa, H. Engineering of a xylose metabolic pathway in Corynebacterium glutamicum. Appl Environ Microbiol 72, 3418-3428 (2006).
19. Inui, M. et al. Metabolic analysis of Corynebacterium glutamicum during lactate and succinate productions under oxygen deprivation conditions. J Mol Microbiol Biotechnol 7, 182-196 (2004).
20. Okino, S. et al. An efficient succinic acid production process in a metabolically engineered Corynebacterium glutamicum strain. Appl Microbiol Biotechnol 81, 459-464 (2008).
21. Okino, S., Suda, M., Fujikura, K., Inui, M. & Yukawa, H. Production of D-lactic acid by Corynebacterium glutamicum under oxygen deprivation. Appl Microbiol Biotechnol 78, 449-454 (2008).
22. Vertès, A.A., Inui, M. & Yukawa, H. Alternative technologies for biotechnological fuel ethanol manufacturing. J Chem Technol Biotechnol 82, 693-697 (2007).

See more of this Session: Developments in Biomass to Fuels
See more of this Group/Topical: Topical 9: Waste & Biomass Valorization

(62b) Evaluation of Pretreatment Methods In the Production of Ethanol From Cattail Leaves

Elijah Wreh, Biological Sciences, St. Cloud State Graduate School, Champlin, MN

Abstract

Previous research in this lab indicated that cattails are a potential source of biomass for the production of cellulosic ethanol since their carbohydrate composition is comparable to that of other plants being considered for biofuel production. To further test their viability, we tested various pretreatment methods on dried cattail leaves. Before polysaccharides in plants can be enzymatically hydrolyzed to fermentable sugars, the plant material must be pretreated to render the polysaccharides accessible to the enzymes. The purpose of this project has been to compare the efficiency of sulfuric acid and ammonia pretreatment methods in preparing cattail biomass for ethanol production. In this project, dried, powdered cattail leaves were pretreated either by autoclaving them with 2% sulfuric acid for one hour or by incubating them overnight at 40 C° in 15 % aqueous ammonia. Samples of the dried, pretreated solid were treated with cellulase and -glucosidase for 48 hours. To compare the efficiency of these pretreatment methods, glucose liberated in these samples was measured by a glucose oxidase assay. It was found that more glucose was recovered in the enzymatic hydrolysis (step two) than in the pretreatment step (step one.) In step two, more glucose was liberated from biomass pretreated with ammonia than from biomass pretreated with sulfuric acid. However, more glucose was recovered from in step one by sulfuric acid pretreatment. Overall 27.8 % of the starting biomass was recovered as glucose with ammonia pretreatment compared to 11.7 % for sulfuric acid pretreatment. Interestingly, 22.3% of the starting biomass was recovered as glucose when no pretreatment was used.

See more of this Session: Developments in Biomass to Fuels
See more of this Group/Topical: Topical 9: Waste & Biomass Valorization

HEMICELLULOSE AND LIGNIN RECOVERY FOR AMMONIUM HYDROXIDE PRETREATMENT PROCESSES

*STEVEN R. SHERMAN, JACOB A. MORRIS, JOHN H. GOODELL**
Savannah River National Laboratory, Aiken, SC, USA
**Adrian College, Adrian, MI, USA*

Introduction

Pretreatment is a step performed in fermentation-based processes that convert lignocellulosic materials into fuels and chemicals. During pretreatment, size-reduced (granulated, chopped, ground, etc.) biomass is subject to physical or chemical steps to make it more amenable to hydrolysis into sugars [1]. While physical pretreatment methods operate by creating more biomass surface area on a microscopic level, chemical pretreatment operates by removing or disrupting the lignin structure that hinders access to cellulosic fibers in the biomass.

Since 2008, Savannah River National Laboratory (SRNL) and Clemson University have been funded by the U.S. Department of Energy, Office of Energy Efficiency and Renewable Energy (DOE-EERE), to develop a fermentation-based process for converting switchgrass, sorghum, and wood into ethanol and other biofuels. The process backbone consists of the following steps: size reduction, pre-treatment, hydrolysis, fermentation, and product purification. The Soaking in Aqueous Ammonia (SAA) pretreatment method [2–4] was initially selected for this work due to prior experience with the method, and it has since proven effective at pretreating these feedstock materials. Application of this method dissolves between 40-60% of lignin, $<$ 20% of hemicellulose, and $<$ 5% of the cellulose from the pretreated biomass. Although the amount of lignin dissolved by this method may be less than can be achieved using other methods, this amount of lignin dissolution has proven sufficient to allow for greater than 80% hydrolysis of pretreated biomass into into sugars when commercially available cellulase mixtures are used to perform the hydrolysis.

As this work proceeds, we have begun to appreciate that the economics of the fermentation-based process may be improved by generating multiple products from the same feed material. During the pretreatment step, lignin is removed from the biomass. This lignin may be recovered from the pretreatment fluid and used as a fuel or a feedstock material higher-value chemicals. Following pretreatment, cellulose is hydrolyzed into glucose (a C-6 sugar), and hemicellulose is hydrolyzed into xylose (a C-5 sugar) and other sugars. In the fermentor, C5-sugars such as xylose are not easily metabolized by natural yeast. As a result, process throughput is reduced and the break-even price of the ethanol product is increased. Although C5-sugar utilization could be improved by using a genetically modified yeast or alternative organism, ethanol has a relatively low market price, and the C5-sugars might be better used to make products that have a higher market value than ethanol.

SRNL recently developed a process for separating and isolating lignin from ammonium hydroxide pretreatment fluids [5], which allows for the generation of a lignin side stream. Some hemicellulose is also extracted during ammonium hydroxide pretreatment, but it is not recovered by the lignin recovery process. Hemicellulose is better extracted using a separate process step, and hemicellulose extraction methods are described in the literature that might be used to recover hemicellulose as a distinct product [4, 6]. When performed before the hydrolysis step, both extraction steps are categorized as pretreatment.

In this work, the performance of the lignin extraction step and the hemicellulose extraction step are examined separately, and then in combination. When both extractions are performed in sequence, the order of the steps is alternated and compared, so that a recommended sequence can be discovered. The lignin extraction is performed using the SRNL method cited previously. The hemicellulose extraction is performed using a method adapted from [6] that uses $Ca(OH)_2$ solution as an extracting agent. Due to lack of chemical analytical data, the chemical composition of the lignin and hemicellulose samples collected during the course of this work is unknown. Instead, sample color and appearance are used as qualitative indicators of product purity in the tests.

Description of Lignin Recovery Method

Lignin is recovered from the ammonium hydroxide pretreatment fluid by the following procedure, and is described more fully in [5].

- L1: The pretreatment fluid is separated from the pretreated biomass material using filtration.
- L2: The ammonium hydroxide pretreatment solution containing dissolved lignin and other materials is heated to the boiling point in order to evaporate ammonia from solution. Ammonia-rich vapor is allowed to evaporate from the solution until the boiling point of the solution reaches 96-98°C at 101.1 kPa.
- L3: While hot, the solution is titrated with 10% sulfuric acid solution until the color of the solution changes from black to brown and the pH of the solution falls to between 2 and 3. The color change indicates the precipitation of a colloid of lignin particles.
- L4: The titrated solution is heated again to the boiling point, and water vapor is allowed to evaporate from the titrated solution until the volume of liquid has fallen to approximately half the starting volume and a "skin" of brown solid begins to form on the sides and top of the liquid. The formation of the "skin" indicates coagulation of colloidal lignin particles into larger particles that are separable from solution by gravity or filtration.
- L5: Allow the mixture to cool, and separate the lignin particles from the liquid by high-speed centrifugation or filtration using a coarse (>25 μm) filter and then a fine (>4 μm) filter.
- L6: Dry the recovered solids in the open air, or in a vacuum oven at a temperature less than 65°C to avoid conglomeration and glazing of the recovered solid.

Description of Hemicellulose Recovery Method

A hemicellulose extraction method adapted from [6] was used to extract and isolate hemicellulose from untreated and pretreated biomass in this work. The extraction steps are provided below. This method was selected for this work because it was well described in the literature, was free to use because the applicable patents had expired, and could be performed using the same equipment as was used for performing lignin extractions.

- H1: Size-reduced biomass (untreated or pretreated) is combined with 0.25 wt% $Ca(OH)_2$ solution in a 10:1 (volume-to-mass) ratio, and the slurry is mixed and heated for 2 hours at 80°C.

- H2: The slurry is allowed to cool, and then the liquid is separated from the solid material by filtration.
- H3: The recovered filtrate is titrated using 10-20% sulfuric acid until the pH of the filtrate is between 2 and 3. The solution may appear yellow-tinted with a fine suspension of white material after titrating it with sulfuric acid.
- H4: The titrated fluid is mixed with methanol in a volumetric ratio of 1.25:1 (volume-to-volume). This causes hemicellulose to precipitate out of solution. The appearance of the solution changes from pale yellow to milky white.
- H5: The precipitated solid may be allowed to settle and the liquid decanted, or the solid-liquid separation process may be accelerated by high-speed centrifugation or filtration.
- H6: The recovered solid, which is white, is allowed to dry in the open air or in a drying oven at a low temperature to avoid melting or caramelizing the recovered hemicellulose.

Results and Discussion

All tests were performed using 50 g samples of biomass as the starting amount of material. The amount of pretreatment solutions used to extract lignin and hemicellulose from the biomass materials were then adjusted to match the mass of the starting material.

Figures 1 and 2 show the percentage of original lignin recovered from previously untreated biomass. Four different feedstock materials were tested: switchgrass, sweet sorghum, sweetgum, and loblolly pine. The percentage of original lignin recovered is defined as the mass of lignin recovered as a solid lignin product divided by the estimated amount of lignin originally present in the biomass, based on characteristic compositional data provided in [8]. According to Figure 1, the amount of lignin recovered varies with the type of biomass when the pretreatment ratio (volume:mass) is 8:1. At an 8:1 pretreatment ratio, the lignin extraction method worked best on switchgrass, and was the least effective on loblolly pine. When the pretreatment ratio was increased from 8:1 to 16:1, the performance of the lignin extraction method on sweet sorghum and loblolly pine was improved, and, for loblolly pine, became statistically indistinguishable from the results obtained for switchgrass (see Figure 2). The convergence of results when the pretreatment ratio was increased from 8:1 to 16:1 suggests that ammonium hydroxide pretreatment, rather than the lignin recovery process, is responsible for the poorer lignin recovery from sweet sorghum, sweetgum, and loblolly pine. The ammonium hydroxide pretreatment process may be affected by a lignin solubility limit, the composition of the biomass, or, more likely, mass transfer limitations that are alleviated at higher liquid-to-solid ratios. The solid material recovered from all of the biomass samples using the lignin recovery procedure was dark brown and powdery, and matched the appearance and consistency of samples that were chemically analyzed and identified as lignin in previous work [5].

The fraction of original hemicellulose recovered from previously untreated samples of switchgrass and sweet gum is shown in Figure 3. In the figure, only about 20% of the original hemicellulose content in the switchgrass samples was recovered as a solid product, while more than 40% of the original hemicellulose content was recovered from samples of sweet gum. Since the same extraction and hemicellulose recovery procedure was used for both materials, it is likely that the differences in the results are due to mass transfer limitations during the extraction. No additional work was performed to

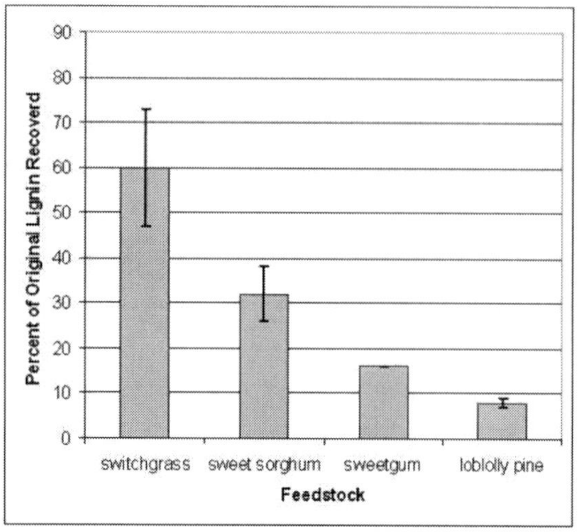

Figure 1: Lignin recovery varies with feedstock at an 8:1 pretreatment ratio.

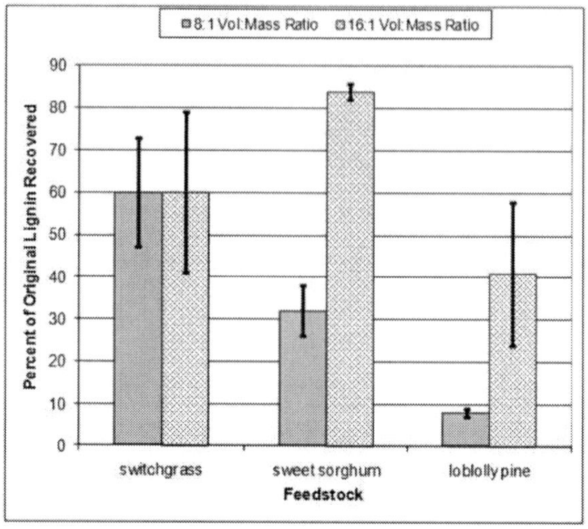

Figure 2: Lignin recovery is improved at a 16:1 pretreatment ratio.

study this question, and more work remains to be done if this hemicellulose extraction method is to be improved and optimized for the biomass feed materials of interest. The hemicellulose recovered from switchgrass had a dull, white, semi-crystalline appearance, while the hemicellulose recovered from sweet gum was bright white and powdery. Both types of hemicellulose turned light brown in color when heated in a drying oven.

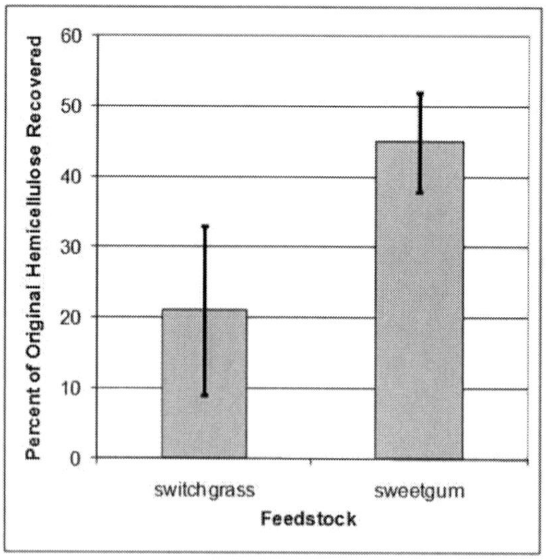

Figure 3: Hemicellulose recovery varies with feedstock.

The switchgrass and sweet gum samples from which hemicellulose had been extracted were then subjected to ammonium hydroxide pretreatment, and lignin was recovered from the ammonium hydroxide pretreatment fluid. The percent of original lignin recovered for these samples is shown in Figure 4. The percentage of original lignin recovered was lower for switchgrass from which hemicellulose had been extracted than for previously untreated switchgrass. For sweetgum, there was no distinguishable difference in the percentage of original lignin recovered, though it is difficult to assign a clear trend for sweetgum when the error bars on the data are so wide.

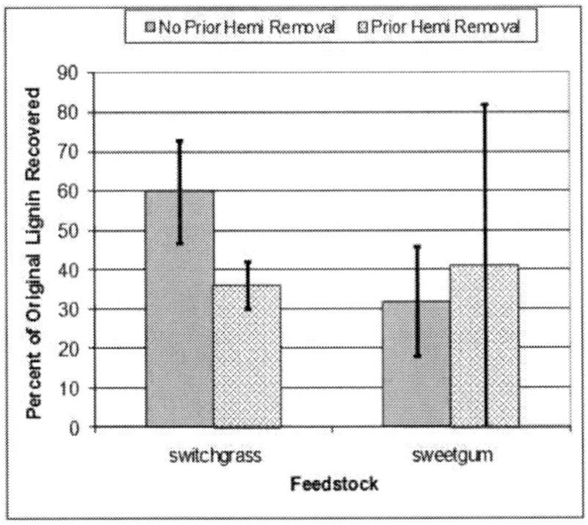

Figure 4: Variation in lignin recovery after some hemicellulose has been removed from the biomass.

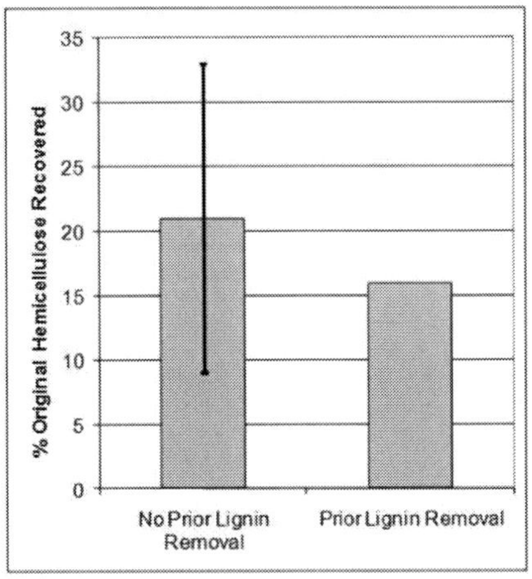

Figure 5: Comparison of hemicellulose extraction results on untreated and NH_4OH pretreated switchgrass.

Figure 5 shows a comparison of the percentage of hemicellulose recovered from untreated switchgrass, and switchgrass that had been pretreated with ammonium hydroxide. Although the chart shows a lower percentage of hemicellulose recovered from switchgrass that had been pretreated, statistically there is no difference between the two cases, and it appear that hemicellulose removal is not affected by ammonium hydroxide pretreatment, at least for the experimental conditions tested here.

Conclusions

The lignin extraction procedure and hemicellulose extraction procedure were shown to be capable of recovering lignin and hemicellulose as separate products from a range of biomass materials (i.e., switchgrass, sweet sorghum, sweetgum, and loblolly pine). The amount of lignin recovered from each type of biomass varied, and it appears that the variation in the amount of lignin recovered was due to mass transfer problems during pretreatment, rather than due to the lignin recovery procedure. The percentage of original hemicellulose recovered from sweetgum was higher than was recovered from switchgrass, which shows that the hemicellulose extraction procedure is affected by biomass type and perhaps mass transfer limitations. The amount of hemicellulose recovered from switchgrass that had been pretreated with ammonium hydroxide was lower than the amount recovered from untreated switchgrass, though this trend was not seen with sweetgum. The amount of lignin recovered from switchgrass that had previously been pretreated with $Ca(OH)_2$ solution was statistically similar to the amount of lignin recovered from untreated switchgrass.

With no clear trends in the data indicating an ideal sequence of extraction steps, it appears that the order in which the hemicellulose and lignin extraction steps is not important. Although there may be some amount of hemicellulose that is extracted along with lignin during ammonium hydroxide pretreatment, apparently the amount of hemicellulose lost from the biomass during the lignin extraction step is small enough that it does not significantly affect the performance of the subsequent hemicellulose extraction step. The reverse sequence, performance of the hemicellulose extraction step before extracting lignin with ammonium hydroxide, shows no effect on the amount of lignin recovered from switchgrass, and so is also a valid extraction sequence.

The variation in the data are wide. More experiments with repeated trials are needed to generate better statistical information before strong conclusions can be drawn on the proper sequencing of the extraction steps. Also, it would be helpful to collect chemical compositional data on the pretreated biomass and recovered products to discern the effects of varying the sequence of extractions on biomass composition and product purity.

References

[1] Y. Sun, J. Cheng, "Hydrolysis of lignocellulosic material for ethanol production: a review", *Bioresource Technology* (2002), 83:1-11.

[2] T.H. Kim, F. Taylor, K.B. Hicks, "Bioethanol production from barley hull using SAA (soaking in aqueous ammonia) pretreatment", *Bioresource Technology* (2008), 99:5694-5702.

[3] T.H. Kim, Y.Y. Lee, "Pretreatment of corn stover by soaking in aqueous ammonia at moderate temperatures", *Applied Biochemistry and Biotechnology* (2007), 137-140:81-92.

[4] T.H. Kim, Y.Y. Lee, "Fractionation of corn stover by hot-water and aqueous ammonia treatment", *Bioresource Technology* (2006), 97:224-232.

[5] S.R. Sherman, J.J. Goodell, C.E. Milliken, J.A. Morris, M.B. Gorensek, "A New Process Developed for Separation of Lignin from Ammonium Hydroxide Pretreatment Fluids", accepted for publication in *Environmental Progress and Sustainable Energy*, December 2010.

[6] M.W. Rutenberg, W. Herbst, "Process for Extraction of Hemicellulose", U.S. Patent 2,801,955, Patented August 6, 1957.

[7] W. Shu-Lai Mok, M.J. Antel Jr., "Uncatalyzed Solvolysis of Whole Biomass Hemicellulose by Hot Compressed Liquid Water", *Industrial Engineering Chemistry Research* (1992), 31:1157-1161.

[8] W.J. Frederick Jr., S.J. Lien, C.E. Courchene, N.A. DeMartini, A.J. Ragauskas, K. Iisa, "Production of ethanol from carbohydrates from loblolly pine: A technical and economic assessment", *Bioresource Technology* (2008), 99:5051-5057.

(62d) Accelerated Hydrolysis of Starch Using Cationic Polyacrylamide for Production of Biofuels

Kendra E. Maxwell, School of Chemical & Biomolecular Engineering Georgia Tech, Georgia Institute of Technology, Atlanta, GA and Sujit Banerjee, School of Chemical & Biomolecular Engineering, Georgia Institute of Technology, Atlanta, GA

Improving the efficiency of the hydrolysis of starch to glucose is key to reducing energy use during the conversion of starch to biofuels. We have found that the addition of positively charged polyacrylamide increases the conversion rate of starch to glucose. The aim of this study is to determine the major factors which affect the mechanism so that optimal production rates are achieved in the presence of the cationic polyelectrolytes. A combination of analytical methods including dynamical light scattering, optical imaging, and uv-vis spectroscopy was used to study polyelectrolyte, starch, and enzyme interaction as a function of process conditions crucial to starch hydrolysis such as composition and temperature. The results indicate that interactions caused by electrostatic forces largely drive the increase in hydrolysis rate. These and related results will be discussed in this presentation.

See more of this Session: Developments in Biomass to Fuels
See more of this Group/Topical: Topical 9: Waste & Biomass Valorization

(62e) Production of Bio-Fuels From Giant Miscanthus

Sanjeev K. Gajjela, Brain Mitchell, Qi Li, El barbary M. Hassan and Philip H. Steele, Forest Products, Mississippi State University, Mississippi State, MS

The energy crop giant miscanthus was pyrolyzed in a laboratory-scale auger reactor at Mississippi State University (MSU). The influence of pyrolysis temperature and biomass pyrolysis time on bio-oil yields is investigated. Pyrolysis temperature of giant miscanthus was varied between 375-450 °C. The highest bio-oil yield of 60 wt% was obtained at 425 °C. Best yield was also a function of biomass feed rate during pyrolysis. We found that maximum yield was attained for an approximate doubling of feedstock feed rate for giant miscanthus as compared to that for pine wood. The reason for this is likely due to differences in particle thickness with wood particles having approximately double or more the thickness of pine wood particles. This apparently allowed the giant miscanthus feedstock particles to pyrolyzed fully at a higher rate. When fed at a the same rate as for pine particles over pyrolysis of the miscanthus occurred as evidenced by lower yields and higher water content of the resulting bio-oil.

Further, raw bio-oil from giant miscanthus in the presence of hydrotreating/hydrocracking catalyst with high temperature and high hydrogen pressure yields hydrocarbons and water. MSU has developed effective hydrotreating (HT) and hydrocracking (HC) catalysts that have successfully produced a hydrocarbon mixture from pine-based bio-oil. These catalysts were applied to the hydroprocessing of giant miscanthus bio-oil. Giant miscanthus bio-oil demonstrated some differences in response to the first-stage HT and second-stage HC treatments as compared to pine bio-oil. These differences will be described in our paper. The physical and chemical properties of giant miscanthus bio-oil and upgraded bio-oil are determined. The fuel-value characteristics of the hydrocarbons produced by two-stage HT and HC will be fully described by analysis by GC/MS and FTIR. Simulated distillation will be performed on the hydrocarbons to describe the boiling range distribution of the hydrocarbon weights produced with comparison of these weights to those of petroleum fuels in the same molecular weight classes.

See more of this Session: Developments in Biomass to Fuels
See more of this Group/Topical: Topical 9: Waste & Biomass Valorization

DEVELOPMENT OF A DETAILED REACTION KINETIC MODEL FOR CELLULOSE FAST PYROLYSIS

Vinu Ravikrishnan and Linda J. Broadbelt, Chemical and Biological Engineering, Northwestern University, Evanston, IL

Fast pyrolysis is one of the potential strategies to convert whole biomass to sustainable transportation fuels. Fast pyrolysis involves heating the biomass in an inert atmosphere to moderate temperatures (400 – 500 °C) in a very short period of time (< 1 s). This leads to a complex network of competing reactions, which results in the formation of bio-oil, non-condensable gaseous species, and solid char. Bio-oil is a mixture of anhydro sugars, phenolic compounds, furan derivatives and low molecular weight oxygenated compounds. The development of a detailed reaction kinetic model to predict the yields of the various pyrolysis products is of utmost importance to understand the relative contributions of the various reaction pathways, and the effect of process parameters like temperature, pressure, and mineral content of biomass.

The existing kinetic models of biomass pyrolysis are global, as they lump the various product species into different fractions like char, condensable and non-condensable gaseous species, thus ignoring the individual identity of the reaction steps that lead to the various products. We have developed a comprehensive mechanistic model for the fast pyrolysis of cellulose, which incorporates the individual steps for the formation of important products like levoglucosan, glycolaldehyde, 5-hydroxymethyl furfural, furfural and formic acid. The underlying mechanism involves the scission of the cellulose chain both at random mid-position and at the chain ends. The heterolytic glucosidic bond scission and the retro-Diels-Alder reactions occur at the mid-chain, while the retro-aldol reaction, 1,2-dehydration, 1,3-dehydration, hydrolysis, and cyclic Grob fragmentation reactions occur at the chain end. The kinetic rate coefficients for most of the above reaction steps were obtained from the literature, based on either experimentally determined Arrhenius parameters for cellulose and model compounds, or theoretically computed values of activation energies using quantum calculations. A computational framework based on continuous distribution kinetics was constructed to solve the kinetic model and predict the time evolution of the product yields. The predictions are compared with experimental data from different sources reported in the literature. The promising results of this study and the applicability of the model for complex biomass systems will be addressed in this presentation.

(93a) Life Cycle Analysis of Biogas-to-Hydrogen Pathways

Jeongwoo Han, Marianne Mintz and Michael Wang, Energy Systems Division, Argonne National Laboratory, Argonne, IL

Methane from organic waste, compromising roughly half of landfill gas (LFG) and about 60% of the gas produced at wastewater treatment plants, farms and other digesters, is a promising renewable source for hydrogen. Until recently, little biogas was recovered in the U.S. to produce electricity and natural gas and none was converted into hydrogen. By 2030, however, California will require that 30 percent of its hydrogen supply come from renewable sources. To examine the impact of such renewable hydrogen standards we conducted a life cycle analysis of various biogas-to-hydrogen pathways by expanding and employing the GREET (Greenhouse Gases, Regulated Emissions, and Energy Use in Transportation) Model, comparing fuel-cycle energy use and greenhouse gas (GHG) emissions from biogas-to-hydrogen pathways to a baseline gasoline pathway as well as other hydrogen pathways. Results show that despite increased total energy use, fossil fuel use and GHG emissions decline for most pathways.

See more of this Session: System Modeling Analyses
See more of this Group/Topical: Topical 9: Waste & Biomass Valorization

(93b) Mathematical Modeling of the Mixing of Municipal Solid Waste (MSW) Particles on the Traveling Grate of a Waste-to-Energy (WTE) Combustion Chamber

Masato R. Nakamura, Marco J. Castaldi and Nickolas J. Themelis, Earth Engineering Center and Department of Earth and Environmental Engineering (Henry Krumb School of Mines), Columbia University, New York, NY

The complex behavior of heterogeneous municipal solid waste (MSW) particles on the traveling grate of a mass-burn waste-to-energy (WTE) combustion chamber affects the steady-state combustion of MSW and often causes an unexpected damage to the walls of a combustion chambers and the traveling grate. In order to understand the parameters that control the combustion processes and designing the grate, we quantified the mixing diffusion coefficient using a stochastic model of MSW particle mixing within a MSW combustion bed. This model was calibrated and validated by means of a full-scale physical model of the Martin reverse acting grate, using tracer particles of sizes ranging from 6 to 22 cm. It was formulated a relationship between the mixing diffusion coefficient and the number of the reciprocating bars in the speed range of 15-90 reciprocations/h of the motion of the reverse acting grate. The ratio of particle diameter to the height of moving bar, d/h, was found to be a major parameter for the mixing diffusion coefficient and the particle residence time at reciprocation speeds exceeding 30 recip./h. Based on these quantitative results and the local MSW particle size distribution, the number of moving bar equipped with the WTE combustion chamber with the grate motion speeds can be designed for optimum operation.

See more of this Session: System Modeling Analyses
See more of this Group/Topical: Topical 9: Waste & Biomass Valorization

(93c) A Model of CO_2-Fed Algal Biofuel Potential In the US

Edward Frank, Energy Systems, Process Technology Research Section, Argonne National Lab, Argonne, IL

A Model of CO2-fed Algal Biofuel Potential in the US

Carbon dioxide is an essential input for some sustainable fuel technologies: The carbon serves as substrate for reduction to a fuel molecule. Large-scale algal biomass production is most often conceived as a pond or photo-bioreactor fed with CO2 from power plant flue gas because mass transfer of CO2 from the atmosphere is too low for the highly productive cultures required for economic viability. We studied the effects of flue gas transport, climate, and power- plant size on algal biomass potential in the US. We report results and concepts that will carry over to other sustainable fuels requiring CO2 inputs.

See more of this Session: System Modeling Analyses
See more of this Group/Topical: Topical 9: Waste & Biomass Valorization

(93d) Optimal Technology Assessment and Product Portfolio Design for Sustainable Biorefineries

Paritosh K. Sharma and Jose Romagnoli, Chemical Engineering, Louisiana State University, Baton Rouge, LA

In recent years there has been a marked surge in the search for alternative sources of energy that wean the world off of dependence on fossil fuels and reduce our carbon foot-print on this world. After a boom in U.S. corn-based ethanol in the early part of the 21st century, the interest has gradually shifted towards more viable sources of biofuels and biochemicals. Cellulosic ethanol, biodiesel, and syngas are examples of such fuels that are extremely attractive owing to the fact that the raw materials can be composed completely of "left-over" wastes of food crops and forest harvests that don't interfere with the human food chain and the natural ecosystem.

Even with increased research emphasis on biofuel and biochemical production technologies, large scale production is still hampered by many factors. Some of these factors that have received considerable interest in recent times include, cultivating more robust feedstock, genetically engineering enzymes and microbes that are more efficient and resistant to poisoning by by-products and improving process technologies that can render large scale production of biofuels a more profitable option for private entities. One area that has not received enough research interest in recent times is strategic modeling and optimization of biorefining enterprises in order to improve strategic decision-making. Such modeling can aid biorefineries to make more informed strategic decisions that will undoubtedly affect the long term profitability, growth, and sustainability of these ventures.

Strategic decisions for nascent biorefineries include product portfolio selection, technology and feedstock selection, and supply chain design and market selection. A sustainable enterprise is often defined as an enterprise that does not have a negative socio-environmental impact on the society. We further refine this definition to encompass not only the ability to positively impact the environment, but also maintaining such an impact through value creation and profitability. An enterprise is defined as being sustainable if it produces goods and services that benefit our environment and is able to preserve such an influence through continued growth. Modeling the strategic decision process of an enterprise that produces renewable-based fuels and chemicals, using engineering and financial tools, can provide valuable insight into the inter-play of technology and process interactions for different product streams, hence providing valuable insights into the correlation between different processing routes which can in turn lead to substantial cost savings for the enterprise. This idea had been motivated by a recent surge in the research area concerning strategic optimization of engineering enterprises and how cost-efficient design of production facilities and supply chains can render an otherwise sluggish enterprise, profitable.

Our work tries to develop a systematic framework, within which strategic decisions a biofuels enterprise makes, can be evaluated. The framework developed is an optimization-based framework which integrates capital budgeting and operational decisions. The decision making process models the actual corporate structure of a process enterprise and is broken down into three concrete steps; the first step is a flexibility evaluation model which provides a preliminary set of portfolio choices of raw materials, suppliers, products, and technologies to the enterprise for detailed economic analysis. The second step evaluates the effects of uncertainty on portfolio choices and provides an incremental strategic plan regarding facility, capacity and portfolio design in order to minimize the downside risks of uncertainty while maintaining their maximum upside. The final step decides on a supply chain and network design, given a process and capacity plan.

The first step in the framework is a deterministic optimization model with the aim of selecting a preliminary feedstock, technology and product portfolio for detailed analysis and evaluating the effect of changing input parameters on the decision variables and the objective function (sensitivity analysis). The model is formulated as an MILP whose outputs include process and capacity design alternatives for different product portfolios. The objective function used in the model is the stakeholder value which is calculated using the free cash flow to firm method. The model formulation includes emissions analysis of process schemes and mandates mitigation of wastes. The constraints and parameters introduced in the model include amongst others, mass and cash balances, production and distribution capacity constraints, and demand constraints. The model decision variables include preliminary process integration schemes and plant capacity design and increments, feedstock selection with seasonal considerations, and semi-annual production, inventory, and sales profiles for each process configuration. Sensitivity analysis is carried out to determine what process configurations offer a judicious mix of design flexibility and profitability. The model acts as a screening model whose results are fed into the second step of the framework for detailed analysis.

The second step of the framework is a real-option based analysis tool formulated as a stochastic MILP problem. Real options analysis is a new paradigm in engineering that borrows techniques from the more established financial options analysis to evaluated assets typically possessed by any manufacturing enterprise. Such analysis has been shown to have practical importance for cases where model parameters have a high degree of uncertainty. A biorefinery of the future will be plagued by uncertainty in raw material prices and availability, product demands and prices, and technological evolution. In such cases, the option to drop (or carry out) a certain action at any point in time has intrinsic value (called real-options value or ROV) and the proposed module seeks to maximize the value of this option. A binomial tree is used to generate scenarios for uncertain parameters and a dynamic programming approach is used to evaluate the portfolio decisions at each time step. Additional constraints include debt balances and obligations, financial ratios and key performance indicators, and raw material availability constraints. The candidate portfolio solutions yielded by the screening model are fed into this module and timing decisions for entry into new feedstock, product and technology markets, raw material supplier selection, and capacity planning are evaluated using this tool.

Following the creation of a plan which includes the optimal process configuration and the corresponding capacity design, the third and final step in the framework is designing the supply chain network for movement of raw materials and finished goods. The structural optimization model mandates the production amounts prescribed by the real-options module leading to fixed revenue streams for the structural analysis module. Consequently, the objective function of the structural analysis module is minimization of total system costs over the optimization time horizon. The model constraints include budgetary constraints, transportation mode-related constraints, and environmental constraints. The outputs for the model include a supply chain configuration, supplier-facility and facility-market matching, transportation mode and network design, and monetary requirements for designing the network.

The potential process scheme that is evaluated using the aforementioned architecture involves simultaneous production of biodiesel and ethanol (thermochemical or biochemical) with a centralized utility facility. Succinic acid is produced as a co-product of cellulosic ethanol with utilization of fermentation-related carbon dioxide emissions. Furthermore glycerol, a by-product of biodiesel, can be utilized in two ways; 1) fermented to produce ethanol, and/or 2) fermented to produce 1,3-propanediol.

The outcomes of our approach for the above configuration are discussed in detail in the paper. The inclusion of environmental and social implications of enterprise activities using life cycle analyses leads us to define our objective function as the stakeholder value. The results show that the despite higher costs, biochemical cellulosic ethanol production has the maximum potential upside and acts as the backbone for the biorefinery. Additionally, an integrated utility generation facility and co-products streams, succinic acid and 1, 3-propanediol, improve the overall profit margins significantly. Furthermore, the co-products provide waste stream utilization opportunities which reduce the overall environmental load of the facility. Sensitivity analysis on the flexibility model indicates that biodiesel production and glycerol fermentation (to produce 1, 3-propanediol) offer tremendous design flexibility as their capacity design depend heavily on the realization of input parameters that affect them. Consequently incremental capacity design for these products can have a significant impact on the profitability of the entire biorefinery.

With mounting competitiveness in the energy sector it will be essential for a biorefining enterprise to assess accurately the options they have for raw material, technology, and product choices. Entry into any market will entail a large capital investment up front and the operation of the enterprise will evolve over time almost certainly with changing market dynamics. Consequently traditional approaches for project appraisal such as the NPV approach will fail to represent an accurate picture since they are static in nature and do not consider uncertainty in model parameters. Hence the systematic approach aforementioned can provide the firm with an essential tool that can be exploited to assess long-term decisions that the enterprise will have to "live-with" once they are made.

See more of this Session: System Modeling Analyses
See more of this Group/Topical: Topical 9: Waste & Biomass Valorization

AUTHOR INDEX